Advanced Optimization for Process Systems Engineering

Based on the author's 40 years of teaching experience, this unique textbook covers both basic and modern advanced concepts of optimization theory and methods for process systems engineers. Topics covered include continuous, discrete, and logic optimization (linear, nonlinear, mixed-integer and generalized disjunctive programming), optimization under uncertainty (stochastic programming and flexibility analysis), and decomposition techniques (Lagrangean and Benders decomposition). Assuming only a basic background in calculus and linear algebra, it enables easy understanding of mathematical reasoning, and numerous examples throughout illustrate key concepts and algorithms. End-of-chapter exercises involving theoretical derivations and small numerical problems, as well as in modeling systems like GAMS, enhance understanding and help put knowledge into practice.

Accompanied by two appendices containing web links to modeling systems and models related to applications in PSE, this is an essential text for single-semester, graduate courses in process systems engineering in departments of chemical engineering.

Ignacio E. Grossmann is the R. R. Dean University Professor of Chemical Engineering at Carnegie Mellon University, and Director of the Center for Advanced Process Decision-making. He is a member of the National Academy of Engineering.

CAMBRIDGE SERIES IN CHEMICAL ENGINEERING

Series Editor

Arvind Varma, *Purdue University*

Editorial Board

Juan de Pablo, *University of Chicago*
Michael Doherty, *University of California-Santa Barbara*
Ignacio Grossmann, *Carnegie Mellon University*
Jim Yang Lee, *National University of Singapore*
Antonios Mikos, *Rice University*

Books in the Series

Daoutidis Baldea, *Dynamics and Nonlinear Control of Integrated Process Systems*

Chamberlin, *Radioactive Aerosols*

Chau, *Process Control: A First Course with Matlab*

Cussler, *Diffusion: Mass Transfer in Fluid Systems, Third Edition*

Moggridge Cussler, *Chemical Product Design, Second Edition*

Schieber De Pablo, *Molecular Engineering Thermodynamics*

Deen, *Introduction to Chemical Engineering Fluid Mechanics*

Denn, *Chemical Engineering: An Introduction*

Denn, *Polymer Melt Processing: Foundations in Fluid Mechanics and Heat Transfer*

Daoutidis Dorfman, *Numerical Methods with Chemical Engineering Applications*

Reimer Duncan, *Chemical Engineering Design and Analysis: An Introduction 2E*

Fan, *Chemical Looping Partial Oxidation Gasification, Reforming, and Chemical Syntheses*

Zhu Fan, *Principles of Gas-Solid Flows*

Fox, *Computational Models for Turbulent Reacting Flows*

Franses, *Thermodynamics with Chemical Engineering Applications*

Leal, *Advanced Transport Phenomena: Fluid Mechanics and Convective Transport Processes*

Shin Lim, *Fed-Batch Cultures: Principles and Applications of Semi-Batch Bioreactors*

Litster, *Design and Processing of Particulate Products*

Fox Marchisio, *Computational Models for Polydisperse Particulate and Multiphase Systems*

Wagner Mewis, *Colloidal Suspension Rheology*

Morbidelli, Gavriilidis, and Varma, *Catalyst Design: Optimal Distribution of Catalyst in Pellets, Reactors, and Membranes*

Nicoud, *Chromatographic Processes*

Terry Noble, *Principles of Chemical Separations with Environmental Applications*

Sandler Orbey, *Modeling Vapor-Liquid Equilibria: Cubic Equations of State and their Mixing Rules*

Petyluk, *Distillation Theory and its Applications to Optimal Design of Separation Units*

Pfister, Nicoud, and Morbidelli, *Continuous Biopharmaceutical Processes: Chromatography, Bioconjugation, and Protein Stability*

Song Ramkrishna, *Cybernetic Modeling for Bioreaction Engineering*

Nott Rao, *An Introduction to Granular Flow*

Russell, Robinson, and Wagner, *Mass and Heat Transfer: Analysis of Mass Contactors and Heat Exchangers*

Schobert, *Chemistry of Fossil Fuels and Biofuels*

Shell, *Thermodynamics and Statistical Mechanics*

Sirkar, *Separation of Molecules, Macromolecules and Particles: Principles, Phenomena and Processes*

Slattery, *Advanced Transport Phenomena*

Varma, Morbidelli, and Wu, *Parametric Sensitivity in Chemical Systems*

Vassiliadis et al., *Optimization for Chemical and Biochemical Engineering*

Weatherley, *Intensification of Liquid–Liquid Processes*

Wolf, Bielser, and Morbidelli, *Perfusion Cell Culture Processes for Biopharmaceuticals*

Zhu, Fan, and Yu, *Dynamics of Multiphase Flows*

Advanced Optimization for Process Systems Engineering

Ignacio E. Grossmann

Carnegie Mellon University, Pennsylvania

CAMBRIDGE
UNIVERSITY PRESS

CAMBRIDGE
UNIVERSITY PRESS

University Printing House, Cambridge CB2 8BS, United Kingdom

One Liberty Plaza, 20th Floor, New York, NY 10006, USA

477 Williamstown Road, Port Melbourne, VIC 3207, Australia

314–321, 3rd Floor, Plot 3, Splendor Forum, Jasola District Centre, New Delhi – 110025, India

79 Anson Road, #06–04/06, Singapore 079906

Cambridge University Press is part of the University of Cambridge.

It furthers the University's mission by disseminating knowledge in the pursuit of education, learning, and research at the highest international levels of excellence.

www.cambridge.org
Information on this title: www.cambridge.org/9781108831659
DOI: 10.1017/9781108917834

First published 2021

Printed in the United Kingdom by TJ Books Limited, Padstow Cornwall, 2021

A catalogue record for this publication is available from the British Library.

ISBN 978-1-108-83165-9 Hardback

Additional resources for this publication at www.cambridge.org/grossmann-resources.

To the memory of my parents

To my wife Blanca and to my children, Claudia, Andrew, and Thomas

To my grandchildren, Oscar, Elena, Joaquin, Lucas, Jamie, and Alex

Contents

Preface

Process systems engineering has emerged as a major discipline in chemical engineering, and is concerned with the systematic analysis and optimization of decision-making processes for the discovery, design, manufacture, and distribution of chemical products. Optimization or, more precisely, mathematical programming, has played a major role in the modeling of decision making in problems related to conceptual design/process synthesis, process control, and process operations. In fact, process systems engineering (PSE) has a rich history in the area of optimization. For instance, some of the first applications of linear programming took place at Gulf Oil in 1952, while the first branch and bound code for solving mixed-integer linear programming was developed at British Petroleum in 1961. The areas of nonlinear and mixed-integer nonlinear programming have been, and continue to be, very active in PSE, as well as logic-based optimization, global optimization, optimization under uncertainty and decomposition methods. In fact, PSE researchers have been making contributions in terms of theory, algorithms, and software, which have had a major impact beyond PSE as they represent fundamental contributions in the area of mathematical programming with applications in areas different than process systems.

The main goal of this book is to provide a unique overview of basic concepts of optimization theory and methods that are relevant to chemical engineers, particularly process systems engineers. The book is unique in that it covers continuous, discrete, and logic optimization, as well as optimization under uncertainty and decomposition techniques. The book is also unique in that it emphasizes several topics related to modeling. The book is the result of having taught over 25 years the course "Advanced Process Systems Engineering" at Carnegie Mellon in which the first part of the course is taught with the first 12 chapters of the book, while the last 3 chapters are taught at the end of the course. The rest of the course is devoted to applications of optimization in process synthesis, and in planning and scheduling of batch and continuous processes.

The book consists of 15 chapters. After the introductory Chapter 1, the book addresses the solution of nonlinear equations through Newton's and quasi-Newton methods in Chapter 2. Chapter 3 introduces basic theoretical optimization concepts, while Chapters 4 and 5 deal with continuous optimization (nonlinear and linear programming). Chapters 6–11 deal with discrete optimization (mixed-integer linear and mixed-integer nonlinear programming, generalized disjunctive programming, and constraint programming). Chapter 12 deals with global optimization, Chapter 13 with Lagrangean decomposition, Chapter 14 with stochastic programming, and Chapter 15 with flexibility analysis. Only knowledge of basic concepts of calculus and linear algebra is assumed. Fourteen of the chapters contain exercises that involve derivations and simple proofs, as well as numerical problems that require the use of the GAMS modeling system. Appendix A describes software for modeling systems optimization algorithms, while Appendix B supplies links to websites that address a variety of problems in PSE.

The author would like to acknowledge the many individuals who made this book possible. I would like to thank the late Professor Sargent who provided me with the basic foundations of optimization, especially through his brilliant course at Imperial College back in 1975. I would like to thank the late Professor Egon Balas whose pioneering work in disjunctive programming has been an inspiration to my students and me for the development of generalized disjunctive programming. I am also most grateful to Professor John Hooker for having introduced me to the area of logic-based optimization. Many thanks to Professor Arthur Westerberg who made possible my hiring at Carnegie Mellon and taught me about the importance of multidisciplinary research. I am also grateful to Professor John Anderson who made possible the teaching of the material in this book through the establishment of the graduate course "Advanced Process Systems Engineering." I am also most grateful to my colleagues from the Center of Advanced Process Decision-making, Larry Biegler, Chrysanthos Gounaris, Nick Sahinidis, Jeff Siirola, and Erik Ydstie, for their help and support. Many thanks to my bright and dedicated former Ph.D. students and research collaborators who have helped to advance the field of optimization. Special thanks to Marco Duran, Gary Kocis, J. Viswanathan, Ramesh Raman, Metin Turkay, Aldo Vecchietti, Sangbum Lee, Nicolas Sawaya, Juan Pablo Ruiz, Francisco Tresplacios, Ignacio Quesada, Juan Zamora, Keshava Halemane, Ross Swaney, Chris Floudas, Stratos Pistikopoulos, and Qi Zhang whose work has been the basis of the various chapters in this book. Thanks also to GAMS Corporation, especially to Alexander Meeraus, for his help in launching our MINLP code DICOPT. I am also indebted to my Ph.D. students David Bernal, Can Li, Qi Chen, Markus Drouven, and Hector Perez, and to Professor Diego Cafaro who helped me to proofread the manuscript and in creating many of the figures. I would also like to acknowledge the Fulbright Award that allowed me a three-month visit to the Universidad Nacional del Litoral/INTEC in Santa Fe, Argentina, that was hosted by Professor Carlos Mendez, and that allowed me to write a large part of the book. I am most indebted to Laura Shaheen and Barbara Carlson for their help and patience in typing and correcting several versions of the manuscript. I am also indebted to Julia Ford, Rachel Norridge and Steve Elliot of Cambridge University Press for their encouragement in writing the book, and their help and advice in the preparation of the manuscript. I also very much hope that this book will be of interest and useful to the Process Systems Engineering community at large, and with whom I have had the privilege of interacting and collaborating during my academic career.

This book is dedicated to my wife Blanca Grossmann, to whom I am most grateful for her love, support, and patience in the writing of this book. The book is also dedicated to the memory of my parents, to my children Claudia, Andrew, and Thomas, and to my grandchildren, Oscar, Elena, Joaquin, Lucas, Jamie, and Alex.

1   Optimization in Process Systems Engineering

1.1   Introduction

This book deals with mathematical programming techniques for process systems engineering (PSE) (Sargent, 2004). The main focus is on the theory, model formulation, and algorithms of mathematical optimization techniques. In this chapter, we provide a general overview of how optimization problems arise in process systems engineering (Biegler et al., 1997). A number of interesting and practical applications are increasingly being tackled in the chemical industry through modern optimization models and methods. The major areas that have been considered include the following:

– process design and synthesis
– planning and scheduling
– process control and operation.

Table 1.1 gives specific areas of application of optimization in process systems engineering (Biegler and Grossmann, 2004; Grossmann and Biegler, 2004). Major optimization techniques include linear programming (LP), nonlinear programming (NLP), mixed-integer linear programming (MILP), mixed-integer nonlinear programming (MINLP), and quadratic programming (QP) (Minoux, 1986).

Figure 1.1 provides the major steps that are involved in the formulation of optimization problems listed in Table 1.1. Given is either a process flowsheet for producing a particular chemical, or a production plan for manufacturing several different products. If we fix the configuration and equipment sizes of a process flowsheet, or the timing of the operations for the production plan, we resort to an analysis of the flowsheet or production plan in order to evaluate the economics (e.g., operating cost), or its performance (e.g., completion time), or its operability (e.g., flexibility or controllability). In order to improve the economics, performance, or operability, we can select new values of parameters that represent degrees of freedom (e.g., pressures, temperatures, reactor volumes, recycle ratios, production rates, etc.). If our aim is to optimize the value of these parameters, mathematically this corresponds to an optimization with continuous variables. We can go a step further and modify the structure of the flowsheet or of the production plan, which can be interpreted as a synthesis step since it gives rise to a new flowsheet structure or to a new production plan in terms of production sequence. If our aim is to optimize these decisions, mathematically this corresponds to an optimization problem with discrete variables.

Table 1.1 Applications of mathematical programming in process systems engineering

	MILP	MINLP	LP	QP	NLP
Design and synthesis					
HENS	X	X	X		X
MENS	X	X	X		X
Separations	X	X			
Reactors		X	X		X
Equipment design		X			X
Flowsheeting		X			X
Operations					
Scheduling	X	X	X		
Supply chain	X	X	X		
Real-time optimization			X	X	X
Linear MPC			X	X	
Nonlinear MPC					X
Hybrid	X				X

Figure 1.1 Major steps in the formulation of optimization problems in PSE.

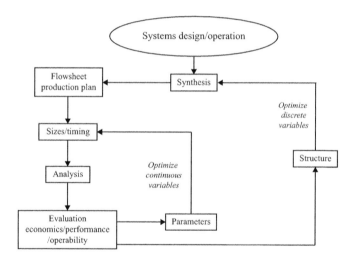

In summary, from Fig. 1.1 we can conclude that the major steps involved in the design/ operation of a process flowsheet or production plan are:

(a) analysis
(b) evaluation
(c) synthesis.

The important point to realize is that the improved design/operation of a flowsheet or production plan can be achieved through the optimization of continuous and discrete variables, which is the main topic of this textbook. Therefore, the outline of this textbook is as follows.

Chapter 2 Solving Nonlinear Equations
Chapter 3 Basic Theoretical Concepts in Optimization
Chapter 4 Nonlinear Programming Algorithms
Chapter 5 Linear Programming
Chapter 6 Mixed-Integer Programming Models
Chapter 7 Systematic Modeling of Constraints with Logic
Chapter 8 Mixed-Integer Linear Programming
Chapter 9 Mixed-Integer Nonlinear Programming
Chapter 10 Generalized Disjunctive Programming
Chapter 11 Constraint Programming
Chapter 12 Nonconvex Optimization
Chapter 13 Lagrangean Decomposition
Chapter 14 Stochastic Programming
Chapter 15 Flexibility Analysis
Appendix A Modeling Systems and Optimization Software
Appendix B Optimization Models for Process Systems Engineering

1.2   Classification of Optimization Models

Optimization problems can generally be formulated with mathematical models that involve continuous and discrete variables (Grossmann, 2002). These variables must be selected to satisfy equations and inequality constraints, while optimizing a given objective function as expressed with the model below:

$$
\begin{aligned}
\min \ & f(x, y) \\
\text{s.t.} \ & h(x, y) = 0 \\
& g(x, y) \leq 0 \\
& x \in X, y \in Y.
\end{aligned}
\tag{1.1}
$$

The continuous variables are represented by x and the discrete variables by y, both with arbitrary dimensionality. The feasible region for the variables x and y is defined by the following constraints: $h(x, y) = 0$, which are equations describing the performance of a system (e.g., mass, energy balances, design equations); and $g(x, y) \leq 0$, which are inequalities that are related to specifications (e.g., minimum purity, maximum throughput). In addition, the variables can be further confined to sets X (typically lower and upper bounds) and Y (specific integers). The former typically specifies the ranges of values for the continuous variables (e.g., positive flows, minimum/maximum pressures for a chemical reactor), while the latter specifies the discrete choices (e.g., only 0-1 choices for the selection or not of a processing unit, or an integer number for number of plates in a distillation column). It should be noted that the model in (1.1) can in practice also be posed as a maximization problem, or with inequalities that must be greater than or equal to zero. The form in (1.1), however, is used for presentation purposes. Also, any problem can be converted into that form because maximizing a function is equivalent to minimizing its negative value. Inequalities that are greater than or equal to zero can also easily be converted to less than or equal to zero.

The mathematical model in (1.1) is technically known as a "mathematical program" and an extensive body of literature exists on this subject (see Williams, 1985, for an introduction to models; and Minoux, 1986, for a general review of methods). From a practical standpoint, the important feature of model (1.1) is that it provides a powerful framework for modeling many optimization problems, including design, scheduling, and control problems. More recently, machine learning models are also being solved through formal optimization methods (Bottou et al., 2018). Depending on the application at hand and the level of detail in the equations, the objective function and constraints can be given in explicit or implicit form. Explicit equations are given by mathematical expressions that are often used for simplified models. In this case, the time to perform an evaluation of the model may be very fast even though it may involve thousands of variables and constraints. Implicit equations arise when detailed and complex calculations must be performed as procedures, as with, for instance, process simulators or differential equation models. In these cases, the dimensionality of the optimization model can be greatly reduced, although the time for performing an evaluation of one trial point may require a significant length of time.

Formulating a given decision problem as a mathematical program in (1.1) requires three major assumptions: (a) the criterion for optimal selection can be expressed through a single objective function, (b) the constraints must be exactly satisfied, and (c) the parameters (input data) are deterministic in nature. These, of course, represent oversimplifications of real-world problems. However, it should be noted that extensions of the model in (1.1) are available for addressing some of these issues. For instance, it is possible to relax the assumptions in (a) and (b) through multiobjective optimization methods (Audet et al., 2008), while the assumption in (c) can be addressed through stochastic optimization methods (Sahinidis, 2004) or multiparametric optimization methods (Pistikopoulos et al., 2007).

If one assumes no special structure to the problem in (1.1) and to its various particular cases (e.g., only discrete or only continuous variables), then direct search techniques are often the easiest methods to apply but also the most time consuming (Rios and Sahinidis, 2013). Here, either a systematic or random selection of trial points is chosen to evaluate and improve the objective function. Satisfaction of constraints can also be accomplished, but frequently with some difficulty (e.g., using penalty functions in which a weighted violation of constraints is added to the original objective function). Perhaps the most popular direct search method that has emerged recently in process systems engineering is simulated annealing. This method is based on analogies with free-energy minimization in statistical mechanics. This method is, in principle, easy to apply to problems with simple constraints, and is likely to find solutions that are close to the global optimum. However, aside from the fact that it often requires many thousands of function evaluations before the likely optimum is found, its performance tends to be highly dependent on the selection of the parameters of the algorithm.

On the other hand, the most prevalent approach that has been taken in optimization is to consider particular problem classes of (1.1) depending on the form of the objective function and constraints with which efficient solution methods can be derived to exploit special structures.

The particular case of (1.1) that is best-known is the linear programming (LP) problem, in which the objective function and all the constraints are linear, and all the variables are continuous (Saigal, 1995; Vanderbei, 2020). As will be discussed in Chapter 5, LP problems have the property that the optimum solution lies at a vertex of the feasible space. Also, any local optimal solution corresponds to the global optimum. These problems have been successfully solved for many years with computer codes that are based on the Simplex algorithm, which is rooted in linear algebra methods. Major changes that have taken place over the past ten years at the level of solution methods are the development of interior-point algorithms that rely on nonlinear transformations and whose computational requirements are theoretically bounded by a polynomial expressed in terms of the problem size (Karmarkar, 1984; Marsten et al., 1990). Interestingly, this property is not shared by the Simplex algorithm, which theoretically may require exponential time. Since this perform-ance is rarely observed in practice, further significant advances have taken place for solving large-scale problems with the Simplex algorithm. With this algorithm, problems with up to 50 000–100 000 constraints can be solved quite efficiently, while interior-point methods tend to perform better in problems with up to 500 000–1000 000 constraints. It should also be pointed out that the mathematical structure of very specialized cases of LP problems (e.g., network flows in assignment or transportation problems) has been greatly exploited, which has made possible the development of codes that can be applied to problems involving millions of variables (Bertsekas, 1998).

The extension of the LP model that involves discrete variables is known as a mixed-integer linear program (MILP) (Nemhauser and Wolsey, 1988; Wolsey, 1998). As discussed in Chapter 6, this model greatly expands the capabilities of formulating real-world problems since one can include logical decisions with 0-1 variables, or account for discrete amounts. As discussed in Chapter 8, the most common method for MILP is the branch and bound search, which consists of solving a subset of LP subproblems while searching within a decision tree of the discrete variables. The other common approach relies on the use of cutting planes that attempt to make the MILP solvable as an LP with the addition of special constraints. Owing to the combinatorial nature that is introduced by the discrete variables in MILP problems, they have proven to be very hard to solve. In fact, theoretically, one can show that this class of problems are NP-hard; that is, there is no known algorithm whose computational requirements do not exceed a polynomial increase in terms of problem size. Nevertheless, recent advances that are based on combining branch and bound methods with cutting planes, and which have been coupled with advances in LP technology, are providing rigorous optimal solutions to problems that were regarded as unsolvable ten years ago.

For the case when all or at least some of the functions are nonlinear, and only continuous variables are involved, problem (1.1) gives rise to nonlinear programming problems (NLP) (Bazaraa et al., 2006; Biegler, 2010). For the case when the objective and constraint functions are differentiable, local optima can be defined by optimality conditions known as the Karush–Kuhn–Tucker (KKT) conditions, as discussed in Chapter 3. These are perhaps the most common type of models that arise in process systems engineering. While ten years ago, problems involving 1000 variables for NLP were regarded as being large, nowadays the solution of problems with up to one million variables is quite common. Reduced-gradient, successive-quadratic programming and interior-point methods, which can be derived by applying Newton's method to the Karush–Kuhn–Tucker conditions, have emerged as the major algorithms for NLP, as will be discussed in Chapter 4. The reduced-gradient method is, in general, better suited for problems with mostly linear constraints, successive-quadratic programming tends to be the method of choice for highly nonlinear problems, while the interior-point method is best suited for very large-scale NLP problems. A limitation of these methods is that they are only guaranteed to converge to a local optimum. For problems that involve a convex objective function and a convex feasible region, this is not a difficulty since these exhibit only one local optimum, which therefore corresponds to the global optimum. In practice, proving convexity in a nonlinear problem is often not possible, and therefore finding any local optimum is often regarded as a satisfactory solution, especially if this yields a significant improvement. On the other hand, there are applications in which finding the global optimum to nonconvex problems is a major issue. Over the past few years, significant progress has been made in developing rigorous methods for globally optimizing problems with special structures (e.g., bilinear functions), as will be discussed in Chapter 12 (Horst and Tuy, 1996; Tawarmalani and Sahinidis, 2004).

The extension of nonlinear programming for handling discrete variables yields a mixed-integer nonlinear programming (MINLP) problem that in its general form is identical to problem (1.1) (Grossmann, 2002; Trespalacios and Grossmann, 2014). MINLP problems were regarded as essentially unsolvable twenty years ago. Algorithms such as the outer-approximation method and extensions of the generalized Benders decomposition method, which are discussed in Chapter 9, have emerged as major methods. These methods assume differentiability in the functions, and consist of solving an alternating sequence of NLP subproblems and MILP master problems. The former optimizes the continuous variables, while the latter optimizes the discrete variables. As in the NLP case, global-optimum solutions can be guaranteed only for convex problems. Solution of problems with typically up to 1000 binary variables and 100 000 continuous variables and constraints has been reported with these methods. Major difficulties encountered in MINLP include those encountered in MILP (combinatorial nature requiring large computations) and in NLP (nonconvexities yielding local solutions).

It should also be noted that all the above methods assume that the problem in (1.1) is expressed through algebraic equations. Very often, however, these models may involve differential equations as constraints. This gives rise to problems known as optimal control problems or the optimization of differential algebraic systems of equations. The major approach that has emerged here is to approximate the differential equations by algebraic equations that yield nonlinear programming problems (Biegler, 2010). The alternate approach is to solve the differential model in a routine that is then treated by the optimizer as a procedure (or implicit function) (Vassiliadis et al., 1994).

Finally, many optimization problems require anticipating the effect of uncertain parameters (Sahinidis, 2004). The most general form of this problem is known as stochastic programming (Birge and Louveaux, 2011), discussed in Chapter 14, in which the basic idea is to optimize over a set of scenarios that arise from discretizing probability distribution functions. Since these problems can give rise to very large-scale problems they are usually solved with decomposition methods. In particular, Benders decomposition is usually used for two-stage stochastic programming models (stage 1: here and now decisions; stage 2: wait and see decisions), while Lagrangean decomposition, which is covered in Chapter 13, is used for multistage stochastic programs by dualizing the so-called "nonanticipativity" constraints. Robust optimization is the other major approach for optimization under uncertainty (Bertsimas et al., 2011), but it is covered only briefly in Chapter 14 as a semi-infinite programming problem. Finally, Chapter 15 examines flexibility analysis, which deals with the problem of determining feasibility in the space of uncertain parameters, while accounting for the fact that corrective action can be taken to regain feasibility.

1.3   Outline of the Book

Having given a general overview of optimization in this chapter, we now provide a general outline of the book to explain the rationale of the order of the chapters, which in part is somewhat nonconventional.

First, we cover Newton's method for nonlinear equation solving in Chapter 2, as it is the basis of deriving the different nonlinear programming algorithms. We also discuss in that chapter quasi-Newton methods that approximate the Jacobian matrix. In Chapter 3 we provide basic concepts on convexity and necessary and sufficient conditions for optimality for unconstrained, equality constrained, and both equality and inequality constrained problems, which give rise to the Karush–Kuhn–Tucker (KKT) conditions. In Chapter 4, we discuss nonlinear programming algorithms to emphasize the fact that NLP algorithms represent the application of Newton's method to the Karush–Kuhn–Tucker conditions. In Chapter 5, we provide some theoretical properties of LP models, and use them as a basis for deriving the Simplex algorithms for which we view it as a special case of nonlinear programming by making connections with the Karush–Kuhn–Tucker conditions and the reduced-gradient method covered in Chapter 4.

Second, we introduce, in Chapter 6, modeling of mixed-integer linear programming problems, stressing the importance of linear 0-1 inequalities. We also introduce several classical combinatorial problems: assignment, facility location, knapsack, set covering, and traveling salesman. Next, in Chapter 7, we introduce the use of propositional logic as a higher-level modeling framework from which linear 0-1 inequalities can be systematically derived. We also introduce disjunctions on continuous constraints, and its two major reformulations as mixed-integer linear constraints big-M and hull reformulations.

Third, we cover in Chapter 8, the major methods for solving mixed-integer linear programs: branch and bound, and cutting planes. In Chapter 9 we describe methods for solving mixed-integer nonlinear programming models, namely branch and bound, outer-approximation (OA) and generalized Benders decomposition (GDB). Although GBD was developed before outer approximation, we show it is easier to derive GDB from OA, and hence the proposed order.

Fourth, we introduce, in Chapter 10, the concept of generalized disjunctive programming, as a high-level representation of optimization problems in terms of continuous and Boolean variables with algebraic, logic, and disjunctive constraints. Aside from showing that the GDP can be used as a modeling framework to derive algebraic mixed-integer linear and nonlinear problems, we also briefly describe disjunctive branch and bound and logic-based outer-approximation algorithms.

Finally, the last five chapters deal with more specialized topics. Chapter 11 describes constraint programming, an alternative logic-based optimization modeling and solution approach based on concepts of logic inference. Chapter 12 discusses global optimization of nonconvex NLP and MINLP models, stressing the use of convex envelopes within spatial branch and bound methods. Chapter 13 describes Lagrangean relaxation and Lagrangean decomposition methods, Chapter 14 describes methods aimed at solving mixed-integer linear stochastic programming models, and Chapter 15 describes formulations and solution methods for flexibility analysis which are relevant for establishing feasibility of constraints under uncertainty with recourse actions.

Appendix A provides information and links to modeling systems such as GAMS, AIMMS, AMPL, Pyomo, and Julia, and to optimization software for LP, NLP, MILP, MINLP, and GDP problems. Appendix B provides links to libraries of problems that contain many application problems in process systems engineering.

2 Solving Nonlinear Equations

2.1 Process Modeling Approaches

In order to perform the analysis of a process system there are two major approaches: equation oriented and sequential modular (Westerberg et al., 1979). Examples of computer software implementing these approaches are gPROMS for the former, and Aspen Plus for the latter.

The basic idea in the equation-oriented approach is to define explicitly all the variables x and assemble all the corresponding equations describing a process or system. Mathematically, this is represented by the system of equations

$$f(x) = 0, \tag{2.1}$$

where f is the vector of equations and x is the vector of variables. Assuming that the equations are independent, the dimensionality of both is the same, i.e. $\dim(f) = \dim(x)$, meaning there are no degrees of freedom. In contrast, in the sequential-modular approach the equations are implemented in a set of modules like the ones presented in Fig. 2.1.

Given an input to module 1 and a guess of the recycle variables x, modules 1, 2, and 3 are executed in sequence so that, given their inputs, the outputs are computed. The execution of module 3 yields the calculated value in the recycle stream, which is denoted by the implicit function $g(x)$. In order to converge to the solution, it is required that the following equation holds true:

$$x = g(x). \tag{2.2}$$

This equation is also known as the fixed-point problem. The existence of the solution to (2.2) is established through the contraction mapping theorem (Ortega and Rheinboldt, 1970). It should be noted that (2.2) can be rearranged in the form of (2.1) by defining $f(x)$ as follows:

$$f(x) = x - g(x). \tag{2.3}$$

Thus, we will consider the solution of problem (2.1) as it represents both modeling approaches. We should note that usually the existence of the solution of (2.1) is established on physical grounds.

2.2 Newton's Method

A general solution method for solving the system of equations in (2.1) is Newton's method (Ortega and Rheinboldt, 1970; Sargent, 1975). Below we present a brief derivation of this method.

Figure 2.1 Example of execution of blocks in the sequential-modular approach.

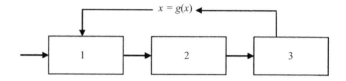

Consider the linear expansion of $f(x)$ at x^k

$$f\left(x^k + p^k\right) = f\left(x^k\right) + J^k p^k,$$ (2.4)

where the computed step p^k is given by

$$p^k = x^{k+1} - x^k$$ (2.5)

and the Jacobian matrix $J^k = J^k\left(x^k\right)$, by the matrix of partial derivatives, is

$$J^k = \left(\frac{\partial f}{\partial x}\right)_{x^k} = \begin{bmatrix} \dfrac{\partial f_1}{\partial x_1} & \dfrac{\partial f_1}{\partial x_2} & \cdots & \dfrac{\partial f_1}{\partial x_n} \\ \dfrac{\partial f_2}{\partial x_1} & \dfrac{\partial f_2}{\partial x_2} & \cdots & \\ \dfrac{\partial f_n}{\partial x_1} & & \cdots & \dfrac{\partial f_n}{\partial x_n} \end{bmatrix}.$$ (2.6)

Notice that each row of J^k represents the gradient of the function $f_i(x)$, i.e. the ith element of the vector $f(x)$. Assuming the solution at $f\left(x^k + p^k\right) = 0$, this gives rise from (2.4) the following system of linear equations,

$$J^k p^k = -f\left(x^k\right),$$ (2.7)

which in turn yields the predicted step

$$p^k = -\left(J^k\right)^{-1} f\left(x^k\right),$$ (2.8)

and hence the new predicted points x^{k+1} are given by

$$x^{k+1} = x^k - \left(J^k\right)^{-1} f\left(x^k\right) \quad k = 0, 1, 2, \ldots,$$ (2.9)

where x^0 is the initial guess. Note that, because the inverse of the Jacobian $\left(J^k\right)^{-1}$ destroys the sparsity of the Jacobian, it is preferable to solve the linear equations in (2.7) in the computation of the step p^k. Hence, the main steps of Newton's algorithm are as follows.

Newton's Algorithm

(1) Guess x^0, set $k = 0$, and select tolerances $\varepsilon_1, \varepsilon_2$.
(2) Calculate $f\left(x^k\right), J^k = J^k\left(x^k\right)$.
(3) Solve linear equations $J^k p^k = -f\left(x^k\right)$ (sparse matrix).

(4) Set $x^{k+1} = x^k + p^k$.
(5) Check convergence with both the error function φ and the error in the difference of the predicted and current point, that is,

$$\varphi = \frac{1}{2} f^T(x^k) f(x^k) \leq \varepsilon_1, \quad \left\| x^{k+1} - x^k \right\| \leq \varepsilon_2.$$

If any of the above inequalities is not satisfied, set $k = k + 1$, and return to step (2).

We should note that in step (2) the derivatives in the Jacobian matrix can be computed analytically or by exact differentiation (Griewank, 2000; Griewank and Walther, 2008).

Some of the major properties of Newton's method include the following (Sargent, 1975).

(1) Second-order convergence.
(2) Sufficient-conditions convergence:
    (a) $f(x)$ must be differentiable,
    (b) $\det(J^k) \neq 0 \forall x$, i.e. Jacobian must be nonsingular,
    (c) initial guess x^0 must be close to the solution (Kantorovich, 1948).
(3) Descent property

$$\varphi(x^k + \alpha p^k) - \varphi(x^k) < 0 \quad \text{for} \quad \alpha \to 0^+.$$

The descent property suggests that p^k be used as a search direction so as to ensure the decrease in the error function φ. The common procedure is as follows

(a) Set $\alpha = 1$.
(b) Calculate $x^{k+1} = x^k + \alpha p^k$.
(c) If $\varphi(x^{k+1}) < \varphi(x^k)$ go to step (5) of Newton's method. Otherwise, set $\alpha^{new} = \theta \alpha^{old}$, where θ is a fractional value (e.g., $\theta = 0.3$), and return to (b).

As an example, let us assume the full predicted step with $\alpha = 1$ leads to an increase of φ from 50 to 80. In that case the full step is not accepted and reduced to $\alpha = 0.3$. Assume this leads to a decrease of φ from 50 down to 45, in which case the step that is taken is with $\alpha = 0.3$.

2.3   Quasi-Newton Methods

The basic idea of quasi-Newton methods (Sargent, 1975; Nocedal, 1980) is to avoid the explicit calculation of the Jacobian by approximating the Jacobian or its inverse, J^k or $(J^k)^{-1}$, with function values. An excellent review on quasi-Newton methods can be found in (Nocedal, 1980). In this section we will focus on Broyden's method (Broyden, 1965) as an example of a quasi-Newton method.

The basic idea in Broyden's method is to generate an explicit update for the Jacobian and successively apply Newton's method. To generate the update, assume that the new Jacobian J^{k+1} can be expressed in terms of the current Jacobian J^k as follows:

$$J^{k+1} = J^k + y^k \left(z^k \right)^T, \tag{2.10}$$

where y^k, z^k are vectors of dimensionality n. Note that $y^k \left(z^k \right)^T$ yields a rank-1 matrix; therefore, Broyden's update formula in (2.10) is known as a rank-1 update formula. The vectors y^k, z^k involve $2n$ unknown elements that can be determined as follows.

First, we consider a given Jacobian J^k for which the step $p^{k+1} = x^{k+1} - x^k$ predicts the function change

$$q^{k+1} = f \left(x^{k+1} \right) - f \left(x^k \right). \tag{2.11}$$

Assuming a first-order expansion as in Newton's method, $f \left(x^{k+1} \right) = f \left(x^k \right) + J^k p^{k+1}$, neglecting the second-order terms, we can rearrange Newton's equation as follows:

$$q^{k+1} = J^k p^{k+1}. \tag{2.12}$$

In order to obtain the new Jacobian, J^{k+1}, we assume the two following conditions to determine the vectors y^k, z^k.

(a) For the step p^{k+1}, $J^{k+1} p^{k+1}$ predicts the same difference of functions q^{k+1} as in (2.12), that is:

$$q^{k+1} = J^{k+1} p^{k+1}, \tag{2.13}$$

which defines n conditions.

(b) For an orthogonal direction r, i.e., $\left(p^{k+1} \right)^T r = 0$, the Jacobians J^k and J^{k+1} predict the same change in the function value, that is,

$$J^{k+1} r = J^k r, \tag{2.14}$$

which defines n additional conditions. Using this along with (2.13) and (2.14) we can calculate y^k, z^k.

We first postmultiply (2.10) by the orthogonal vector r,

$$J^{k+1} r = J^k r + y^k \left(z^k \right)^T r. \tag{2.15}$$

From (2.14), it follows that $y^k \left(z^k \right)^T r = 0$. This can be accomplished if we select $z^k = p^{k+1}$. Next, we postmultiply (2.10) by p^{k+1},

$$J^{k+1} p^{k+1} = J^k p^{k+1} + y^k \left(p^{k+1} \right)^T p^{k+1}. \tag{2.16}$$

From (2.13), we can express (2.16) as

$$q^{k+1} = J^k p^{k+1} + y^k \left(p^{k+1} \right)^T p^{k+1}, \tag{2.17}$$

from which we can solve for y^k,

$$y^k = \frac{q^{k+1} - J^k p^{k+1}}{(p^{k+1})^T p^{k+1}}. \tag{2.18}$$

Therefore, Broyden's update for the Jacobian J^k is given by

$$J^{k+1} = J^k + \frac{(q^{k+1} - J^k p^{k+1})(p^{k+1})^T}{(p^{k+1})^T p^{k+1}}. \tag{2.19}$$

To update the inverse of the Jacobian, $(J^k)^{-1}$, we make use of the Sherman–Morrison formula (Sherman, 1949),

$$(A + xy^T)^{-1} = A^{-1} - \frac{A^{-1}xy^T A^{-1}}{1 + y^T A^{-1}x}, \tag{2.20}$$

which leads to Broyden's update of the inverse of the Jacobian $(J^k)^{-1}$

$$(J^{k+1})^{-1} = (J^k)^{-1} + \frac{(p^{k+1} - (J^k)^{-1}q^{k+1})(p^{k+1})^T (J^k)^{-1}}{(p^{k+1})^T (J^k)^{-1} q^{k+1}}. \tag{2.21}$$

The steps in Broyden's algorithm using the update in (2.21) for the inverse of the Jacobian, are as follows

(1) (a) Set $k = 0$, guess $x^0, \varepsilon_1, \varepsilon_2$.
   (b) Calculate $q(x^0)$, estimate J^0 by finite differences.
      Calculate: $(J^0)^{-1}$.
(2) Calculate $p^{k+1} = -(J^k)^{-1} f(x^k)$.
   Set $x^{k+1} = x^k + p^{k+1}$.
(3) Calculate $f(x^k), q^{k+1} = f(x^{k+1}) - f(x^k)$.
(4) Check convergence as in Newton's method with both the error function φ and the error in the difference of the predicted and current point. If convergence is not achieved, update $(J^{k+1})^{-1}$, set $k = k + 1$, and return to step (2).

We should note that, in (2.21), we need to store the matrix $(J^k)^{-1}$ which destroys the sparsity that might be present in the original Jacobian. Also, as in Newton's method, we can introduce and adjust the step size α in step (2). Finally, the initial Jacobian is usually obtained by perturbation, in which the partial derivatives are estimated from the approximation

$$\frac{\partial f_i}{\partial x_j} \cong \frac{f_i(x^k + he_j) - f_i(x^h)}{h}, \tag{2.22}$$

Figure 2.2 Perturbation in two dimensions for given point x^0.

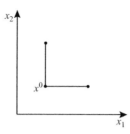

where h is the size of the perturbation and e_j is the unit vector along the jth coordinate:

$$e_j = \begin{pmatrix} 0 \\ 0 \\ 1 \\ 0 \end{pmatrix}. \tag{2.23}$$

Note that (2.21) requires $n + 1$ function evaluations. Also, we present in Fig. 2.2 a two-dimensional example of a perturbation in the x_1 and x_2 direction for a given initial point x^0. Notice that three function evaluations are required in this case.

EXERCISES

2.1 Specify three cases for which Newton's method may fail to converge when solving the square system of nonlinear equations $f(x) = 0$.

2.2 Show that for a sufficiently small positive step size α, the step predicted by Newton's method exhibits the descent property as given by $\varphi(x^k + \alpha p^k) - \varphi(x^k) < 0$, where $\varphi(x) = \frac{1}{2} f^T(x) f(x)$, and x^k is any point for which $f(x^k) \neq 0$ and the Jacobian J^k exists.

2.3 The update formula for the Jacobian in Broyden's method is given by:

$$J^{k+1} = J^k + \frac{\left(q^{k+1} - J^k p^{k+1}\right)\left(p^{k+1}\right)^T}{\left(p^{k+1}\right)^T p^{k+1}},$$

where J^k is the Jacobian matrix ($n \times n$-matrix)

$$q^{k+1} = f\left(x^{k+1}\right) - f\left(x^k\right) \quad (n\text{-vector})$$

$$p^{k+1} = x^{k+1} - x^k \quad (n\text{-vector}).$$

Using the Sherman–Morrison formula

$$\left(A + xy^T\right)^{-1} = A^{-1} - \frac{A^{-1}xy^T A^{-1}}{1 + y^T A^{-1}x},$$

derive the update formula for the inverse of the Jacobian in Broyden's method.

2.4 Assume Broyden's update formula is applied to the linear equations, $Ax = b$, where A is a square matrix:

$$J^{k+1} = J^k + \frac{\left(q^{k+1} - J^k p^{k+1}\right)\left(p^{k+1}\right)^T}{\left(p^{k+1}\right)^T p^{k+1}}.$$

If you start with the exact Jacobian for the initial guess, show that Broyden's formula predicts at the first iteration that same Jacobian. Does this result require the matrix A to be nonsingular?

2.5 The system of equations

$$f_1 = 2x_1^2 + x_2^2 - 5 = 0$$
$$f_2 = x_1 + 2x_2 - 3 = 0$$

is to be solved with the starting point $x_1 = 1.5$, $x_2 = 1.0$, and with the criterion of convergence $\frac{1}{2}f^T f \le 10^{-8}$, using the two following methods:

(a) Newton's method with unit step size $(\alpha = 1)$,
(b) Broyden's method with the update formula for the inverse of the Jacobian, using unit step size $(\alpha = 1)$ and estimating the initial Jacobian J^0 by using a perturbation of $h = 10^{-4}$ in the variables.

Indicate in each case the number of iterations required to achieve the desired convergence.

3   Basic Theoretical Concepts in Optimization

3.1   Basic Formulations

In this section we introduce basic concepts of nonlinear constrained optimization as a preamble of Section 3.4 on optimality conditions.

We first consider the model of a process given by the set of m equations h in n continuous real-valued variables,

$$h(x) = 0. \tag{3.1}$$

The equations in (3.1) typically represent equations that describe the performance of a system, say mass and energy balance equations and sizing equations. If $n > m$ this implies that there are degrees of freedom in the set of equations (3.1). If we define the number of degrees of freedom to be $df = n - m$, then, if $df > 0$, there is a finite number of degrees of freedom, and therefore the model equations are amenable to optimization. Specifically, we want to select the degree of freedom to optimize a given objective function. Say our goal is to minimize the objective function represented by the scalar objective $f(x)$ in n variables, $f : R^n \rightarrow R^1$. Aside from having to satisfy the equations in (3.1) the variables x may be subject to a set of r inequalities that are represented by the constraints $g(x) \leq 0$, $g : R^n \rightarrow R^r$. Physically these may represent ranges of choices of the variables x, or simply nonnegativity conditions on these variables.

If we gather the objective function, model equations, and inequalities, the continuous optimization problem can be expressed mathematically as follows (Biegler, 2010):

$$
\begin{aligned}
\min \quad & f(x) \\
\text{s.t.} \quad & h(x) = 0 \\
& g(x) \leq 0 \\
& x \in R^n.
\end{aligned}
\tag{3.2}
$$

Assuming that at least some of the functions in (3.2) are nonlinear, problem (3.2) corresponds to a nonlinear programming (NLP) problem. If all the functions are linear then the problem corresponds to a linear programming (LP) problem. In both cases, they are generally known as mathematical programming problems.

We can make a few remarks on problem (3.2).

(a) It can be easily proved that $\min f(x)$ is equivalent to $\max -f(x)$. Figure 3.1 provides a geometrical illustration of this property.

Figure 3.1 Illustration of the relation between minimization and maximization of objectives.

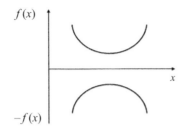

(b) Inequalities can always be rearranged in the form of $g(x) \leq 0$. As a simple example consider the inequality

$$x_1 + 2x_2 \geq 3, \tag{3.3}$$

which can be rearranged as follows:

$$-x_1 - 2x_2 + 3 \leq 0. \tag{3.4}$$

(c) If a subset of the variables are discrete, $y \in Y^q$, say integer or 0-1, this then gives rise to the following problem, which corresponds to a mixed-integer nonlinear programming (MINLP) problem:

$$
\begin{aligned}
\min \quad & f(x, y) \\
\text{s.t.} \quad & h(x, y) = 0 \\
& g(x, y) \leq 0 \\
& x \in R^n, \ y \in Y^q.
\end{aligned}
\tag{3.5}
$$

(d) Both problems (3.2) and (3.5) are declarative models, in the sense that in both the formulations the conceptual problem is to find the optimum value of x^* in (3.2) and optimal values (x^*, y^*) in problem (3.5) irrespective of how they are to be solved numerically.

3.2 Nonlinear Programming Example

Consider the optimal design of the cooler shown in Fig. 3.2, in which a hot stream is to be cooled from 550 K down to 320 K using cooling water that is available at 300 K. Its outlet temperature t_w is constrained not to be higher than 330 K.

Given the overall heat transfer coefficient U, the unit costs c_A, c_W for the area and cooling water, respectively, the problem consists of finding the design of the cooler (area A, heat load Q, outlet cooling water t_w) that minimizes the total cost, namely area and cooling water.

Without even writing the optimization model, we can qualitatively describe the problem through Fig. 3.3, in which we plot the total cost versus the flowrate of cooling water W. As

Figure 3.2 Optimal design of a cooler.

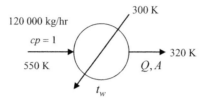

Figure 3.3 Optimal trade-off between area and cooling water cost.

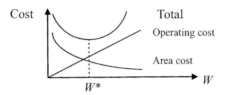

can be seen, the cost of cooling water (operating cost) increases linearly with the flowrate of cooling water. On the other hand, the cost of the area (investment cost) decreases as we increase the flowrate W. This follows from the fact that on the one hand at smaller flowrates W, the outlet temperature increases, leading to smaller driving forces, and hence larger areas. On the other hand, at larger flowrates, the driving force increases leading to smaller areas. Adding both curves gives a total cost function whose minimum value is given at the value W^*. The solution to this optimization problem, which establishes an optimum trade-off between investment and operating cost can be formulated as the following NLP problem.

In order to formulate the optimization model, let Q, A be the heat load and area, and let W and t_w be the water flowrate and its outlet temperature such that $t_w \leq 330$ K. The corresponding nonlinear programming model can be formulated in the form of a general NLP model in (3.5) as follows:

$$\min \quad C = c_A A + c_W W$$

$$\text{s.t.} \quad Q - 120\,000(550 - 320) = 0$$

$$Q - W(t_w - 300) = 0$$

$$Q - UA \frac{(550 - t_w) - (320 - 300)}{\ln \dfrac{550 - t_w}{320 - 300}} = 0$$

$$t_w - 330 \leq 0$$

$$300 - t_w \leq 0$$

$$-Q \leq 0$$

$$-A \leq 0$$

$$-W \leq 0.$$

$$(3.6)$$

Compared to (3.5) the variables x correspond to Q, A, W, t_w. The three equations for the heat balances and area sizing correspond to the equations $h(x) = 0$, while the bounds on t_w, $300 \le t_w \le 330$, and $Q, A, W \ge 0$, have been rearranged in the form of the inequalities $g(x) \le 0$.

We should note that (3.6) corresponds to a declarative model, in which the objective function and constraints as in (3.5) are explicitly declared without regard to how they will be solved. Also note that the NLP has one degree of freedom as there are four variables, Q, A, W, t_w, and three equations. We should also note that, in software codes for NLP, problem bounds can be specified separately from the inequalities $g(x) \le 0$.

3.3   Basic Concepts

This section introduces basic key theoretical concepts in optimization (Sargent, 1975; Minoux, 1986).

Definition 3.1 Feasible region $F = \{x | x \in R^n, h(x) = 0, g(x) \le 0\}$.

Note that the feasible region F is given by the intersection of the constraints, and is equivalent to applying the logic AND ($\wedge$) operator to the set of equalities and inequalities (e.g., see Fig. 3.4).

Definition 3.2 F is convex if and only if, for any $x^1, x^2 \in F$,

$$x = \alpha x^1 + (1 - \alpha)x^2 \in F, \quad \forall \alpha \in [0, 1].$$

Figures 3.5 and 3.6 illustrate examples of convex and nonconvex feasible regions.
The following sufficient condition can be established for convexity of a feasible region.

Property 3.1 If $h(x) = 0$ is linear and $g(x) \le 0$ involves convex functions, then F is a convex feasible region.

Definition 3.3 $f(x)$ is convex if and only if, for every $x^1, x^2 \in R^n$

$$f\left(\alpha x^1 + (1 - \alpha)x^2\right) \le \alpha f\left(x^1\right) + (1 - \alpha)f\left(x^2\right) \ \forall \alpha \in [0, 1].$$

Figure 3.4 Geometric representation of feasible region for three inequalities $g(x) \le 0$ and one equality $h(x) = 0$. Arrows point to infeasible directions.

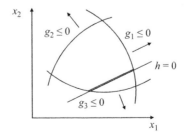

Figure 3.5 Example of convex feasible region.

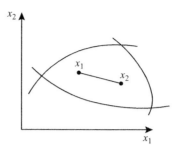

Figure 3.6 Example of nonconvex feasible region.

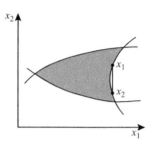

Figure 3.7 Examples of convex and nonconvex (concave) functions.

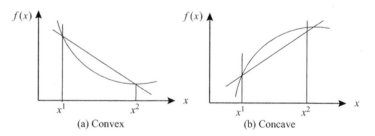

Figure 3.7 illustrates geometrically a convex function and a concave function that is nonconvex.

Definition 3.4

(a) Local minimum. A point $\hat{x} \in F$ corresponds to a local minimum if and only if, $f(x) \geq f(\hat{x})$ for $0 < \|x - \hat{x}\| < \delta$, $x \in F$, for δ arbitrarily small and positive.

(b) Global minimum. A point $\hat{x} \in F$ corresponds to a global minimum if and only if,

$$f(x) \geq f(\hat{x}) \quad \forall x \in F.$$

The above are said to be strong local and global optima if the strict inequality applies ($>$) (see Fig. 3.8). If the equality applies, the above are said to be weak local and global optima.

Figure 3.8 Examples of strong local and weak local minima.

3.4 Optimality Conditions

3.4.1 Unconstrained Optimization

Consider the unconstrained optimization problem:

$$\min_{x \in R^n} f(x) \tag{3.7}$$

where we assume that $f(x)$ is a continuous, differentiable function. A necessary condition for the point $\hat{x}$ to be an extremum of problem (3.7) is given by the following theorem (Bazaraa et al., 2006).

Theorem 3.1 *If $f(x)$ has an extremum in $\hat{x}$, then $\hat{x}$ is a stationary point, which implies*

$$\nabla f(\hat{x}) = 0.$$

The first-order optimality condition of (3.7) (setting the gradient of f to zero ($\nabla f(\hat{x}) = 0$) can be posed as the system of n equations in n unknowns,

$$\begin{bmatrix} \dfrac{\partial f}{\partial x_1} \\ \vdots \\ \dfrac{\partial f}{\partial x_n} \end{bmatrix} = \begin{bmatrix} 0 \\ \vdots \\ 0 \end{bmatrix}. \tag{3.8}$$

The system of equations in (3.8) can be solved with Newton's method.

For the sufficient condition of point $\hat{x}$ to be an extremum, consider the second-order expansion of $f(x)$ at the point $\hat{x}$,

$$f(\hat{x} + \Delta x) = f(\hat{x}) + \nabla f(\hat{x})^T \Delta x + \frac{1}{2} \Delta x^T H \Delta x, \tag{3.9}$$

where H is the Hessian matrix of second derivatives evaluated at $\hat{x}$ (e.g. x_1, x_2),

$$H = \begin{bmatrix} \dfrac{\partial^2 f}{\partial x_1{}^2} & \dfrac{\partial^2 f}{\partial x_1 \partial x_2} \\ \dfrac{\partial^2 f}{\partial x_2 \partial x_1} & \dfrac{\partial^2 f}{\partial x_2{}^2} \end{bmatrix}. \tag{3.10}$$

This matrix is symmetric since the order of differentiation is immaterial for Lipschitz continuous functions. Since at the extremum $\nabla f(\hat{x}) = 0$, this implies from (3.9) that

$$\Delta x^T H \Delta x > 0 \quad \forall \Delta x \neq 0. \tag{3.11}$$

A Hessian matrix H satisfying the above condition is said to be positive definite (strictly positive definite for >0, and semipositive definite for ≥ 0). Since in general it is not trivial to establish the positive definiteness of a Hessian matrix, we can exploit the following properties, which are based on the analysis of the signs of the eigenvalues or on the convexity of the functions (Wilkinson, 1965):

(a) H is positive definite if and only if the eigenvalues $\lambda_i > 0$, $i = 1, 2, \ldots, n$,
(b) if $f(x)$ is strictly convex, then H is positive definite.

We should also note that if $f(x)$ is strictly convex and it has a stationary point $\hat{x}$, then $\hat{x}$ is a unique local minimum.

3.4.2 Constrained Optimization (Equalities)

We address in this section optimization problems with equality constraints, which can be formulated as follows:

$$\min \ f(x)$$
$$\text{s.t.} \ h(x) = 0 \tag{3.12}$$
$$x \in R^n,$$

where we assume that $f(x), h(x)$ are continuous, differentiable functions in the continuous variables x.

The necessary conditions for an extremum at a point $\hat{x}$ in (3.6) are given by the following theorem

Theorem 3.2 *If $f(x)$ has a constrained extremum at $\hat{x}$ such that $h_j(\hat{x}) = 0$ $j = 1, \ldots, m$, then the gradients $\nabla f(\hat{x}) = 0$, $\nabla h_j(\hat{x})$ are linearly dependent, that is,*

$$\lambda_0 \nabla f(\hat{x}) + \sum_{j=1}^{m} \lambda_j \nabla h_j(\hat{x}) = 0 \tag{3.13}$$

where λ_0, λ_j are scalars.

Proof Consider the $m + 1$ equations, where y is the value of the function $f(x)$,

$$\begin{aligned} f(x) &= y \\ h_j(x) &= 0 \quad j = 1 \ldots m \end{aligned} \tag{3.14}$$

A sufficient condition for the existence of a solution around $\hat{x}$ for any y close to the assumed extremum $\hat{y}$, is that the Jacobian of (3.14),

$$J(\hat{x}) = \begin{bmatrix} \nabla f^T(\hat{x}) \\ \nabla h_1^T(\hat{x}) \\ \cdots \\ \nabla h_m^T(\hat{x}) \end{bmatrix}, \tag{3.15}$$

has rank $m + 1$. Since the rank of the Jacobian $J(\hat{x})$ is $m + 1$, this implies that we can find solutions for $y > \hat{y}$, which contradicts the assumption that $\hat{y}$ is an extremum. This then implies that the rank of $J(\hat{x})$ is strictly $< m + 1$. This in turn implies that the rows in $J(\hat{x})$ are linearly independent, namely,

$$\lambda_0 \nabla f(\hat{x}) + \sum_{j=1}^{m} \lambda_j \nabla h_j(\hat{x}) = 0, \tag{3.16}$$

which is identical to (3.13).   ■

We should note that if $\lambda_0 = 1$, then this means that a "constraint qualification" (Bazaraa et al., 2006) holds and, hence,

$$\nabla f(\hat{x}) = - \sum_{j=1}^{m} \lambda_j \nabla h_j(\hat{x}), \tag{3.17}$$

which means that at the extremum $\hat{x}$, the gradient of the objective function can be expressed as a linear combination of the gradients of the equalities. Figure 3.9 presents a geometrical interpretation for the case of two dimensions and a quadratic objective function, where it can be seen that ∇f, and ∇h are co-linear.

Figure 3.10 shows a case where the constraint qualification does not hold true since the gradient of the objective ∇f, cannot be obtained by a finite linear combination of the gradients of the equations ∇h_1, ∇h_2.

Figure 3.9 Minimum for a constrained quadratic function for which the constraint qualification is satisfied.

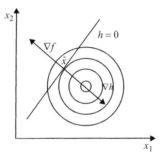

Figure 3.10 Minimum for a constrained quadratic function for which the constraint qualification is not satisfied.

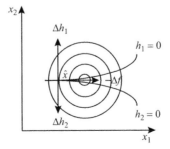

Theorem 3.2 can also be interpreted as follows. If we set $\lambda_0 = 1$, it follows from (3.17) and the condition of feasibility for the equations, that,

$$\nabla f(\hat{x}) = -\sum_{j=1}^{m} \lambda_j \nabla h_j(\hat{x}) = 0 \tag{3.18}$$

$$h_j(\hat{x}) = 0 \quad j = 1, \ldots, m, \tag{3.19}$$

which correspond to the stationary conditions of the Lagrangean function,

$$L = f(x) + \sum_{j=h}^{m} \lambda_j h_j(x) \tag{3.20}$$

since

$$\frac{\partial L}{\partial x} = \nabla f(x) + \sum_{j=1}^{m} \lambda_j \nabla h_j(\hat{x}) = 0 \tag{3.21}$$

$$\frac{\partial L}{\partial \lambda_j} = h_j(\hat{x}) = 0 \quad j = 1, \ldots, m, \tag{3.22}$$

which gives rise to a system of $n + m$ nonlinear equations in the $n + m$ unknowns (x, λ). The nonlinear equations in (3.21) and (3.22) are typically solved with Newton's method.

3.4.3 Constrained Optimization (Inequalities)

We consider the following nonlinear optimization problem with inequalities,

$$\min \ f(x)$$

$$\text{s.t.} \quad g(x) \leq 0 \tag{3.23}$$

$$x \in R^n,$$

where $f(x), g(x)$ are assumed to be continuous, differentiable functions. A necessary condition for the point $\hat{x}$ to be a local minimum of (3.17) is given by the following theorem by Fritz John (1948).

Theorem 3.3 *If $f(x)$ has a constrained minimum at $\hat{x}$ such that $g_j(\hat{x}) \leq 0 \ \hat{x}, j = 1, \ldots, r$, then there exist nonnegative multipliers $\mu_j \geq 0, j = 0, 1, \ldots, r \geq$ such that*

$$\mu_0 \nabla f(\hat{x}) + \sum_{j=1}^{r} \mu_j \nabla g_j(\hat{x}) = 0 \tag{3.24}$$

$$\mu_j g_j(\hat{x}) = 0 \quad j = 1, \ldots, r. \tag{3.25}$$

The first equation defines the condition of linear dependence of the gradients of the objective function and inequality constraints, while the second equation establishes complementary conditions between the multipliers μ_j and the corresponding inequalities $g_j(\hat{x}) \leq 0$ evaluated at the constrained minimizer $\hat{x}$. These imply that if $\mu_j > 0$, then $g_j(\hat{x}) = 0$, i.e. the inequality g_j is active. If $\mu_j = 0$, then $g_j(\hat{x}) \leq 0$; i.e. the inequality is inactive if $g_j(\hat{x}) < 0$.

The proof of Theorem 3.3 can be obtained by applying Gordan's Transposition Theorem (Gordan, 1873). Rather than presenting the proof, we illustrate this theorem with the example in Fig. 3.11 that involves two variables, x_1 and x_2, and the three inequalities $g_1(x) \leq 0, g_2(x) \leq 0, g_3(x) \leq 0$. Notice that at the optimum solution, $\hat{x}$, there are two active constraints, $g_1(\hat{x}) = 0, g_2(\hat{x}) = 0$, and one inactive inequality $g_3(\hat{x}) < 0$. Furthermore, we can set $\mu_0 = 1$, since a constraint qualification is assumed to hold true. Therefore, from Theorem 3.3 the following conditions hold true at $\hat{x}$, namely linear dependence of gradients of the objective and active inequalities, and corresponding complementarity conditions

$$\nabla f + \mu_1 \nabla g_1 + \mu_2 \nabla g_2 = 0$$
$$g_1 = 0, \quad g_2 = 0, \quad \mu_1, \mu_2 \geq 0$$
$$g_3 < 0, \quad \mu_3 = 0 \tag{3.26}$$
$$\mu_1 g_1 = 0, \ \mu_2 g_2 = 0, \ \mu_3 g_3 = 0.$$

Referring to Fig. 3.11 it is interesting to note that the gradient of the objective ∇f can be viewed as given by a linear combination of the gradients of the active constraints $\nabla g_1, \nabla g_2$.

Also, note that the multipliers in (3.26) have an interesting interpretation. If we set $\mu_0 = 1$, then (3.21) reduces to

$$\nabla f(\hat{x}) + \sum_{j=1}^{r} \mu_j \nabla g_j(\hat{x}) = 0 \tag{3.27}$$

with $\mu_j \geq 0$. Let us consider the direction vector p shown in Fig. 3.11. We can project the gradients of the objective and inequalities by considering the following scalar products, which yield the changes δf in the objective, and δg_j in the constraints,

$$p^T \nabla f(\hat{x}) = \delta f, \quad p^T \nabla g_j(\hat{x}) = \delta g_j. \tag{3.28}$$

Figure 3.11 Two-dimensional example with two active inequalities.

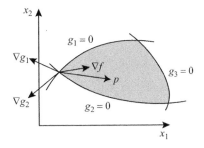

To relate these changes, we first premultiply (3.27) by the vector p,

$$p^T \nabla f(\hat{x}) + \sum_{j=1}^{m} \mu_j p^T \nabla g_j(\hat{x}) = 0. \tag{3.29}$$

By substituting (3.28) we obtain

$$\delta f + \sum_{j=1}^{m} \mu_j \delta g_j = 0. \tag{3.30}$$

If we now set $\delta g_j = 0 \, \forall j \neq i$, (3.30) reduces to

$$\delta f + \mu_i \delta g_i = 0, \tag{3.31}$$

from which it follows that,

$$\mu_i = - \left(\frac{\delta f}{\delta g_i} \right)_{\substack{\delta g_j = 0 \\ j \neq i}}. \tag{3.32}$$

Equation (3.32) states that when $\delta g_i \rightarrow 0$, the multiplier μ_i represents the rate of decrease of the objective $f(\hat{x})$ with a small perturbation of the inequality $g_i(\hat{x})$, at the optimum solution $\hat{x}$. Furthermore, we can interpret (3.32) in terms of the sign of the multipliers. Specifically, consider an infeasible move $\delta g_i > 0$ at the active inequality $g_j(\hat{x}) = 0$.

(a) For $\mu_i > 0$, (3.32) implies $\delta f < 0$; in other words a local decrease of the objective for an infeasible perturbation of inequality constraint i;
(b) for $\mu_i < 0$ (3.32) implies $\delta f > 0$; in other words, a local increase of the objective for an infeasible perturbation of inequality constraint i.

As we will see in the next section, the sign of the multipliers can be used within an active strategy for solving nonlinear programming problems.

3.4.4 Nonlinear Programming Problem

We consider the more general form of nonlinear programming (NLP) problems that involve both equality and inequality constraints, $h(x) = 0$, $g(x) \leq 0$ (h: m-vector, g: r-vector):

$$\begin{aligned} \min \quad & f(x) \\ \text{s.t.} \quad & h(x) = 0 \\ & g(x) \leq 0 \\ & x \in R^n. \end{aligned} \tag{3.33}$$

The necessary conditions for a constrained minimum at the point $\hat{x}$ in the NLP in (3.33) are given by the Karush–Kuhn–Tucker (KKT) conditions (Karush, 1939; Kuhn and Tucker, 1951), which result from combining the results for the NLP with only equalities and only inequalities, and assuming a constraint qualification. The KKT conditions are given by the three following conditions.

(1) Linear dependence of gradients

$$\nabla f(\hat{x}) + \sum_{j=1}^{m} \lambda_j \nabla h_j(\hat{x}) + \sum_{j=1}^{r} \mu_j \nabla g_j(\hat{x}) = 0. \tag{3.34}$$

(2) Constraint feasibility

$$h_j(\hat{x}) = 0, \quad j = 1, \ldots, m \quad g_j(\hat{x}) \leq 0 \quad j = 1, \ldots, r. \tag{3.35}$$

(3) Complementarity conditions

$$\begin{aligned} \mu_j g_j(\hat{x}) &= 0 \quad j = 1, \ldots, r \\ \mu_j &\geq 0 \quad j = 1, \ldots, r. \end{aligned} \tag{3.36}$$

The following properties hold for the case of convex NLPs (convex objective and convex feasible region):

Property 3.2 If $f(x)$ is convex, the feasible region F is convex and nonempty, then if there exists a local minimum at $\hat{x}$, then:

(a) $\hat{x}$ corresponds to a global minimum,
(b) the constraint qualification is satisfied,
(c) the KKT conditions are necessary and sufficient for a global minimum.

The main implication of the above properties for convex NLPs is that there is no need to analyze second-order information of the Hessian of the Lagrangean of (3.33), $\nabla_{xx} L(\hat{x}, \mu, \lambda)$, i.e. determine whether the Hessian is positive definite to define a local minimum.

In contrast to the unconstrained case (3.7) or the equality-constrained case (3.12), where one can solve for the optimality conditions as a system of equations (e.g., with Newton's method), solving for the KKT conditions (3.34)–(3.36) is not straightforward due to the complementarity conditions (3.36), which give rise to nonconvex equations.

The simplest method for finding a point that satisfies the KKT conditions is an iterative active-strategy method in which a subset of inequalities are iteratively forced to become active until they all satisfy the complementarity conditions and attain the correct sign of the multipliers. The procedure outlined below provides the major steps involved in the active-set strategy.

3.4.5 Active-Set Strategy Procedure for Determining a Karush–Kuhn–Tucker Point (Sargent, 1975)

Let $J_1 = \{j | g_j(x) = 0\}$ be the index of active inequalities.

(1) Assume no active inequalities. Set $j_1 = \emptyset$, $u_j = 0$, $j = 1, 2, \ldots, r$.
(2) Formulate KKT equations (3.31)–(3.33) and solve for x, λ_j, $j = 1, 2, \ldots, m$, and μ_j, $j \in J_1$,

$$\nabla f(x) + \sum_{j=1}^{m} \lambda_j \nabla h_j(x) + \sum_{j \in J_1} \mu_j \nabla g_j(x) = 0$$

$$h_j(x) = 0, j = 1, \ldots, m$$

$$g_j(x) = 0, j \in J_1.$$

(3) If at the solution $\hat{x}$, $g_j(\hat{x}) \leq 0$ and $\mu_j \geq 0$, $j = 1, 2, \ldots, r$, then STOP. Solution found.

(4) If any $g_j(\hat{x}) > 0$ (violated) and/or any $\mu_j < 0$ (wrong sign), then do the following.

    (a) Drop one active constraint with wrong sign for μ_j (largest magnitude).

    (b) Add to J_1 violated constraints $g_j(x) > 0$ to make them active (maximum of n constraints).

    (c) Return to step (2).

We should note that the main goal in the active-set strategy is to identify the correct set of active constraints. In step (1), we can start with the initial guess x^0 and with a given set of active constraints rather than assuming that they are all inactive. Also, the reason we drop only one constraint in step (4a) is that the multiplier corresponds to the partial derivative; that is,

$$\mu_j = -\partial f / \partial g_j \quad \text{if } \mu_j < 0 \Rightarrow \text{ for } \partial g_j < 0, \partial f < 0.$$

We next present an example to illustrate the application of the active strategy for finding a KKT point in the following NLP problem, which in fact corresponds to a quadratic program since the objective is quadratic and the constraints are linear:

$$\begin{aligned} \min \quad & f(x) = \tfrac{1}{2}\left(x_1^2 + x_2^2\right) - 3x_1 - x_2 \\ \text{s.t.} \quad & g_1 = -x_1 + x_2 \leq 0 \\ & g_2 = x_1 - \tfrac{1}{2}x_2 - 2 \leq 0 \\ & g_3 = -x_2 \leq 0. \end{aligned} \tag{3.37}$$

Figure 3.12 shows the feasible region and contours of the objective function, where by inspection we can see that the minimum solution lies at the point $x_1 = 2.6$, $x_2 = 1.2$, in which the inequality g_2 is active.

To apply the active-set strategy, we first determine the gradients of the objective function and three inequalities:

$$\nabla f = \begin{bmatrix} x_1 - 3 \\ x_2 - 1 \end{bmatrix} \quad \nabla g_1 = \begin{bmatrix} -1 \\ 1 \end{bmatrix} \quad \nabla g_2 = \begin{bmatrix} 1 \\ -1/2 \end{bmatrix} \quad \nabla g_3 = \begin{bmatrix} 0 \\ -1 \end{bmatrix}.$$

We follow the steps of the active-set strategy.

First iteration

(1) Set $J_1 = \emptyset$, $\mu_1 = \mu_2 = \mu_3 = 0$.

(2) Since there are no active inequalities, (3.34) reduces to the stationary condition of $f(x)$:

$$\nabla f(x) = 0,$$

Figure 3.12 Feasible region and contours of objective for example problem.

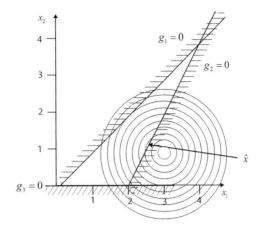

which in turn give rise to the equations:

$$x_1 - 3 = 0$$
$$x_2 - 1 = 0$$

whose solution yields $x_1 = 3$, $x_2 = 1$.

(3) Substituting these values in the three inequalities yields:

$$g_1 = -2 < 0 \quad g_2 = \tfrac{1}{2} > 0 \quad g_3 = -1 < 0.$$

(4) Make g_2 active since it is violated; i.e., set $J_1 = \{2\}$, meaning we treat it as the equation $g_2 = x_1 - \tfrac{1}{2}x_2 - 2 = 0$.

Second iteration

(2) Since g_2 is active, the stationary conditions of the Lagrangean are as follows:

$$\nabla f(x) + \mu_2 \nabla g_2(x) = 0$$
$$g_2(x) = 0,$$

which in turn corresponds to a linear system of three equations in three unknowns x_1, x_2, μ_2,

$$\begin{bmatrix} x_1 - 3 \\ x_2 - 1 \end{bmatrix} + \mu_2 \begin{bmatrix} 1 \\ -\tfrac{1}{2} \end{bmatrix} = \begin{bmatrix} 0 \\ 0 \end{bmatrix}$$
$$x_1 - \tfrac{1}{2}x_2 - 2 = 0,$$

whose solution yields $x_1 = 2.6$, $x_2 = 1.2$ $\mu_2 = 0.4$.

(3) Substituting these values in the three inequalities yields:

$$g_1 = -1.4 < 0 \quad g_2 = 0 \quad g_3 - 1.2 < 0$$
$$\mu = 0 \quad \mu_2 = 0.4 \quad \mu_3 = 0.$$

Since the three inequality constraints are satisfied and the multipliers have the correct sign to satisfy the complementary conditions, we can stop and conclude that the point $x_1 = 2.6$, $x_2 = 1.2$ with objective value $f = -4.9$ corresponds to a KKT point of the quadratic program (3.37).

It is interesting to note that, at the KKT point $x_1 = 2.6$, $x_2 = 1.2$, linear dependence of gradients holds since for:

$$\nabla f = \begin{bmatrix} -0.4 \\ 0.2 \end{bmatrix}, \quad \nabla g_2 = \begin{bmatrix} 1 \\ -\frac{1}{2} \end{bmatrix}, \quad \mu_2 = 0.4.$$

We can easily verify the linear dependence of the gradients of the objective and active inequality, that is:

$$\begin{bmatrix} -0.4 \\ 0.2 \end{bmatrix} + 0.4 \begin{bmatrix} 1 \\ -0.5 \end{bmatrix} = \begin{bmatrix} 0 \\ 0 \end{bmatrix}.$$

We can also establish that the KKT point found corresponds to a minimum that is unique (i.e. global minimum). If we determine the Hessian matrix and substitute the corresponding values this yields the identity matrix

$$H = \begin{bmatrix} \dfrac{\partial^2 f}{\partial x_1{}^2} & \dfrac{\partial^2 f}{\partial x_1 \partial x_2} \\ \dfrac{\partial^2 f}{\partial x_2 \partial x_1} & \dfrac{\partial^2 f}{\partial x_2{}^2} \end{bmatrix} = \begin{bmatrix} 1 & 0 \\ 0 & 1 \end{bmatrix}.$$

The eigenvalues of the identity matrix can be computed by setting to zero the determinant of the identity matrix minus the eigenvalues in the diagonal, that is,

$$\begin{vmatrix} 1 - \lambda & 0 \\ 0 & 1 - \lambda \end{vmatrix} = (1 - \lambda)^2 = 0.$$

The solution of the quadratic polynomial yields $\lambda_1 = \lambda_2 = 1$. Since both eigenvalues are strictly positive, it follows that the objective function $f(x)$ is strictly convex. Furthermore, since the inequality constraints are linear, the feasible region is convex. It then follows that the quadratic programming problem (3.37) is convex. Hence it has a unique optimum, and the KKT conditions are both necessary and sufficient.

EXERCISES

3.1 Show that if $f(x)$ is a convex function on R^n, then a local minimum of $f(x)$ is also a global minimum of $f(x)$ on R^n.

3.2 Determine whether or not the following quadratic function is convex:

$$f(x_1, x_2) = 2x_1 + x_2 - x_1{}^2 + 2x_1 x_2 - x_2{}^2.$$

3.3 Given is a set of linear equalities $Ax = b$ and a set of convex nonlinear inequalities $g(x) < 0$, where A is an $m \times n$ matrix and x an n-vector ($m < n$). Assuming that the corresponding feasible region is nonempty, show that the region is convex.

3.4 Assume that you are given a quasiconvex function

$$g(x). \tag{3.38}$$

Show that its inequality in the form of $g(x) \le 0$ defines a convex feasible region.

3.5 Show that a Hessian matrix which is positive definite has positive eigenvalues.

3.6 Consider the nonlinear programming problem

$$\begin{aligned} \min \quad & f(x) \\ \text{s.t.} \quad & g_j(x) \le 0 \quad j = 1, \ldots, r \\ & x \in R^n, \end{aligned}$$

where the functions f and g are monotonic in each variable $x_i x_i$ (i.e., $\partial f / \partial x_i$ and $\partial g_j / \partial x_i$ are one-signed if the derivatives are nonzero).

Show that the following hold true (these are the so-called principles of monotonicity analysis).

(a) If a variable x_i is present in the objective and in some of the constraints, there is at least one active constraint involving x_i and whose derivative has an opposite sign from the objective.

(b) If a variable x_i is not present in the objective, it is either involved in at least two active constraints with an opposite sign in the derivatives, or else it is not involved in any active constraints.

3.7 Given the problem

$$\begin{aligned} \min \quad & f(x) = (x_1 - 3)^2 + (x_2 - 4)^2 \\ \text{s.t.} \quad & 2x_1 + x_2 - 6 \le 0 \\ & x_1 - x_2 - 4 \le 0 \\ & x_1 \ge 1.8 \\ & x_2 \ge 0 \end{aligned}$$

do the following.

(a) Plot the contours for $f = 0, 1, 2, 4, 6, 8, 10$, and the feasible region. From inspection, what is the optimal solution?

(b) Solve this problem using an active-constraint strategy. Plot the gradients of the objective and active constraints at the optimum and verify geometrically the Karush–Kuhn–Tucker conditions. Determine also whether the optimal solution is unique.

4   Nonlinear Programming Algorithms

We consider in this chapter algorithms for the solution of NLP problems of the following form:

$$\begin{aligned}
\min \quad & f(x) \\
\text{s.t.} \quad & h(x) = 0 \\
& g(x) \leq 0 \\
& x \in R^n.
\end{aligned} \tag{4.1}$$

The algorithms we will consider include the following.

(a) The successive-quadratic programming (SQP) algorithm that is implemented in the code SNOPT in GAMS.
(b) Reduced-gradient algorithm that is implemented in the code CONOPT in GAMS.
(c) Interior-point method that is implemented in the code IPOPT, which is open source software.

As we will show in the next few sections, the various algorithms apply Newton's method to the KKT conditions in a different way.

4.1   Successive-Quadratic Programming

The successive-quadratic programming algorithm, also known as the SQP algorithm, or the Wilson–Han–Powell algorithm (Wilson, 1963; Han, 1976; Powell, 1978), is based on the idea of applying Newton's method to the KKT conditions of the NLP with equality constraints:

$$\begin{aligned}
\min \quad & f(x) \\
\text{s.t.} \quad & h(x) = 0 \\
& x \in R^n.
\end{aligned} \tag{4.2}$$

Setting up the Lagrangean function of (4.2),

$$L(x, \lambda) = f(x) + h(x)^T \lambda, \tag{4.3}$$

the necessary conditions for a stationary point of the Lagrangean, yield the following:

(a) $\dfrac{\partial L}{\partial x} = \nabla f(x) + \nabla h(x) \lambda = 0$

(b) $\dfrac{\partial L}{\partial \lambda} = h(x) = 0,$

$$\tag{4.4}$$

where $\nabla h = [\nabla h_1, \nabla h_2, \ldots, \nabla h_m]$ is the transpose of the Jacobian. Note that (4.4a) and (4.4b) define $n + m$ equations in $n + m$ unknowns for the variables x and λ. We now represent (4.4a) and (4.4b) by the system of equations

$$\phi(u) = 0. \tag{4.5}$$

Applying Newton's method to (4.5) at a given point $u^i = (x^i, \lambda^i)$,

$$\phi(u^i) + \nabla\phi(u^i)^T (u - u^i) = 0, \tag{4.6}$$

from (4.4a) this implies,

$$\nabla f(x^i) + \nabla h(x^i)\lambda^i + \nabla_{xx} f(x^i)(x - x^i) + (\lambda^i)^T \nabla_{xx} h(x^i)(x - x^i) + \nabla h(x^i)(\lambda - \lambda^i) = 0 \tag{4.7}$$

which can be simplified as

$$\nabla f(x^i) + \nabla h(x^i)\lambda + \nabla_{xx} f(x^i)(x - x^i) + (\lambda^i)^T \nabla_{xx} h(x^i)(x - x^i) = 0, \tag{4.8}$$

where $\nabla_{xx} h = [\nabla_{xx} h_1, \ldots, \nabla_{xx} h_m]$. Note also that the term $(\lambda^i)^T \nabla_{xx} h(x^i)(x - x^i)$ can be written in expanded form as $\sum_j \lambda_j^i \frac{\partial^2 h_j}{\partial x^2}(x - x^i)$.

Since the Hessian of the Lagrangean, $\nabla_{xx} L(x^i, \lambda^i) = \nabla_{xx} f(x^i) + (\lambda^i)^T \nabla_{xx} h(x^i)$, in (4.8) together with the linearization of the equalities $h(x) = 0$, this yields,

$$\begin{aligned} f(x^i) + \nabla h(x^i)\lambda + \nabla_{xx} L(x^i, \lambda^i)(x - x^i) &= 0 \\ h(x^i) + \nabla h(x^i)^T (x - x^i) &= 0. \end{aligned} \tag{4.9}$$

While (4.9) correspond to the equations for Newton's method applied to (4.4a) and (4.4b), the interesting observation is that they also correspond to the KKT conditions of the quadratic program with the objective function

$$\nabla f(x^i)^T (x - x^i) + \tfrac{1}{2}(x - x^i)^T \nabla_{xx} L(x^i, \lambda^i)(x - x^i) \tag{4.10}$$

and constraints

$$h(x^i) + \nabla h(x^i)^T (x - x^i) = 0. \tag{4.11}$$

That is, we can compute the Newton step $d = x - x^i$ for the system of equations in (4.9) from the following quadratic program (QP):

$$\begin{aligned} \min \quad & \nabla f(x^i)^T + \tfrac{1}{2} d^T \nabla_{xx} L(x^i, \lambda^i) d \\ \text{s.t.} \quad & h(x^i) + \nabla h(x^i)^T (x - x^i) d = 0, \end{aligned} \tag{4.12}$$

where the Hessian of the QP is the Hessian of the Lagrangean. The solution of (4.12) yields the Newton step $d = x - x^i$ and the multipliers λ^{i+1}. The new guess for the variables x can then be set to $x^{i+1} = x^i + d$.

One may of course wonder what we gain by solving Newton's method through the QP. Aside from the fact that there are efficient methods for solving QP problems, the extension for handling inequalities becomes trivial. That is, for given x^i, λ^i, μ^i we can set up the QP:

$$\min \quad \nabla f(x^i)^T + \tfrac{1}{2} d^T \nabla_{xx} L(x^i, \lambda^i, \mu^i) d$$
$$\text{s.t.} \quad h(x^i) + \nabla h(x^i)^T d = 0 \tag{4.13}$$
$$g(x^i) + \nabla g(x^i)^T d \leq 0,$$

where the Lagrange function is given by

$$L(x, \lambda, \mu) = f(x) + \sum_{j=1}^{m} \lambda_j h_j(x) + \sum_{j=1}^{r} \mu_j g_j(x).$$

In the implementation of the algorithm, the Hessian $\nabla_{xx} L$ is often approximated with a matrix B^i using the quasi-Newton formula BFGS (Nocedal and Wright, 2006) to force positive definiteness. Also, at each iteration a line search is performed to select the step size α in $x^{i+1} = x^i + \alpha d$ to minimize the Lagrangean. We should also note that the SQP algorithm is an infeasible path algorithm because it simultaneously optimizes the objective while converging the constraints. Furthermore, being based on Newton's method, it has quadratic convergence close to optimum.

In summary, the major steps involved in the SQP algorithm (Wilson–Han–Powell) are as follows.

(1) Set iteration counter $i = 0$; set estimate Hessian of Lagrangean $B^0 = I$; guess x^0:
(2) Evaluate the functions and the gradients.

$$f(x^i), h(x^i), g(x^i), \nabla f(x^i), \nabla h(x^i), \nabla g(x^i).$$

(3) Solve the quadratic program QP to obtain step d, and multipliers λ^{i+1}, μ^{i+1}

$$\min \quad \nabla f(x^i)^T d + \tfrac{1}{2} d^T B^i d$$
$$\text{s.t.} \quad h(x^i) + \nabla h(x^i)^T d = 0 \quad \rightarrow \quad [d, \lambda^{i+1}, \mu^{i+1}]$$
$$g(x^i) + \nabla g(x^i)^T d \leq 0.$$

(4) Check convergence:

$$\text{If } \left\| \nabla L(x^i, \lambda^{i+1}, \mu^{i+1}) \right\|_2 < \varepsilon, \text{ STOP,}$$
where $\nabla L = \nabla f + \nabla h \lambda + \nabla g \mu$.
Else set $x^{i+1} = x^i + \alpha d$ $(\alpha = 1 : \text{Newton})$,
and go to step 5.

(5) Update the Hessian of the Lagrangean (BFGS).
   (a) Determine $s^i = x^{i+1} - x^i$
   (b) Calculate $y^i = \nabla L(x^{i+1}, \lambda^{i+1}, \mu^{i+1}) - \nabla L(x^i, \lambda^i, \mu^i)$ (rank-2 update).
   (c) Set $i = i + 1$, $B^{i+1} = B^i - \dfrac{B^i s^i s^{iT} B^i}{s^{iT} B^i s^i} + \dfrac{y^i y^{iT}}{y^{iT} s^i}$ return to step (2).

4.2   Reduced-Gradient Method

In this section we consider the reduced-gradient method as originally proposed by Murtagh and Saunders (1978, 1982) with the code MINOS, and later extended by Drud (1994) with the code CONOPT.

The basic ideas in the reduced-gradient method are to consider a linear approximation of the constraints and to eliminate variables to reduce dimensionality, and then to apply Newton's method.

Consider the following NLP with linear equalities,

$$\begin{aligned} \min \quad & f(x) \\ \text{s.t.} \quad & Ax = b, \end{aligned} \tag{4.14}$$

where A is an $m \times n$ matrix, $m < n$, and for simplicity in the presentation we do not consider nonnegativity conditions $x \geq 0$. In order to eliminate variables from the linear constraints we partition the variables x as follows:

$$x = \begin{bmatrix} y \\ u \end{bmatrix}, \tag{4.15}$$

where y are the m dependent (i.e. calculated) variables, and u are the $n - m$ independent or superbasic variables (i.e., degrees of freedom). If we now partition the matrix as follows, $A = [B|C]$, where matrix B is an $m \times m$ square matrix, and C is an $m \times (n - m)$ matrix, we can then rewrite the linear constraints as

$$Ax = [B|C] \begin{bmatrix} y \\ u \end{bmatrix} = b. \tag{4.16}$$

Consider the step Δx, such that

$$\Delta x = \begin{bmatrix} \Delta y \\ \Delta u \end{bmatrix} = \begin{bmatrix} y - y^k \\ u - u^k \end{bmatrix} \tag{4.17}$$

satisfies (4.16). We can then rewrite (4.16) at the predicted and current points as

$$\begin{aligned} By + Cu &= b \\ By^k + Cu^k &= b. \end{aligned} \tag{4.18}$$

By subtracting both we obtain

$$B\Delta y + C\Delta u = 0, \tag{4.19}$$

from which it follows that $\Delta y = -B^{-1}C\Delta u$, and hence the predicted step Δx is given by

$$\Delta x = \begin{bmatrix} \Delta y \\ \Delta u \end{bmatrix} = \begin{bmatrix} -B^{-1}C \\ I \end{bmatrix} \Delta u. \tag{4.20}$$

We can also rewrite (4.20) as

$$\Delta x = Z\Delta u, \tag{4.21}$$

where Z, the transformation matrix, is given by

$$Z = \begin{bmatrix} -B^{-1}C \\ I \end{bmatrix}. \tag{4.22}$$

To apply Newton's method to problem (4.14), consider a second-order expansion of $f(x)$ at x^i,

$$f(x) = f(x^i) + \nabla f(x^i)^T \Delta x + \frac{1}{2}\Delta x^T \nabla_{xx} f(x^i)\Delta x. \tag{4.23}$$

If we substitute (4.21) in (4.23), it can be expressed in terms of the independent variables u,

$$f(x) = f(x^i) + \nabla f(x^i)^T Z\Delta u + \frac{1}{2}\Delta u^T Z^T \nabla_{xx} f(x^i) Z\Delta u. \tag{4.24}$$

We define as the reduced gradient the term, $g_R^T = \nabla f(x^i)^T Z$, and as the reduced Hessian, $H_R = Z^T \nabla_{xx} f(x^i) Z$. Hence (4.24) can then be expressed in terms of the reduced gradient and reduced Hessian, namely

$$f(u) = f(x^i) + g_R^T \Delta u + \frac{1}{2}\Delta u^T H_R \Delta u. \tag{4.25}$$

Since a necessary condition for a minimum of $f(x)$ is given by the stationary condition $\frac{\partial f}{\partial \Delta u} = 0$, we can set

$$g_R + H_R \Delta u = 0. \tag{4.26}$$

To determine the Newton step Δu, we rearrange (4.26) as follows,

$$H_R \Delta u = -g_R, \tag{4.27}$$

which gives rise to a linear system of $n - m$ equations in $n - m$ variables to compute the Newton step Δu.

We should note that, in the implementation of the reduced-gradient method, as in the SQP method, the reduced Hessian matrix $Z^T = \nabla_{xx} f(x^i) Z$ is usually estimated with a quasi-Newton formula (e.g., BFGS).

Also, to compute the reduced gradient $g_R^T = \nabla f(x^i)^T Z$, we can partition it into the dependent (y) and independent (u) variables as follows:

$$g_R^T = \begin{bmatrix} \nabla_y f & \nabla_u f \end{bmatrix}^T \begin{bmatrix} -B^{-1}C \\ I \end{bmatrix} = -\nabla_y f^T B^{-1} C + \nabla_u f^T. \tag{4.28}$$

In order to avoid computing the inverse of the matrix B^{-1} we solve the linear system for the multipliers λ from the KKT equation,

$$B^T \lambda = -\nabla_y f. \tag{4.29}$$

Since $(B^T \lambda)^T = \lambda^T B = -\nabla_y f^T$, and by postmultiplying by B^{-1}, the reduced gradient can be obtained from the following equation,

$$g_R^T = \lambda^T C + \nabla_u f^T. \tag{4.30}$$

Finally, for handling nonlinear constraints, we reformulate the NLP in (4.1) in terms of equalities $r(x) = 0$ as follows,

$$\min \quad f(x)$$
$$\text{s.t.} \quad r(x) = 0 \tag{4.31}$$
$$x \geq 0.$$

This reformulation can be obtained by adding slack variables to the inequalities. As a simple example consider the inequality $x_1 + x_2 \leq 1$. By introducing a nonnegative slack variable s, we can rewrite that constraint as the equality $x_1 + x_2 + s = 1$, where the slack variable $s \geq 0$.

To handle the nonlinear inequalities $r(x) = 0$ in (4.31), we define an augmented Lagrangean in an "outer iteration" in which we add a dual term to the objective and where we linearize the equations $r(x) = 0$:

$$(P^i) : \min \quad f(x) + \left(\lambda^i\right)^T \left[r(x) - r\left(x^i\right)\right]$$
$$\text{s.t.} \quad J\left(x^i\right)x = b \tag{4.32}$$

where $J(x^i)$ is the Jacobian matrix of $r(x)$, and λ^i are the multipliers computed at the point x^i.

The procedure then consists of formulating a sequence of (P^i) subproblems in an outer iteration, each of which is solved in an "inner iteration" using the reduced-gradient method, because (4.32) has the form of the linearly constrained problem (4.14). At each outer iteration the new point is obtained by setting $x^{i+1} = x^i + \alpha[x^* - x^i]$, where x^* is the solution (P^i) and α is the step length, which is adjusted to ensure descent of the augmented Lagrange function in (4.32).

In the procedure below, we summarize the main steps of the reduced-gradient method as implemented in the original code MINOS (Murtagh and Saunders, 2003). Part I refers to the inner iterations for the linearly constrained problem, while Part II refers to the outer iterations.

(I) Inner iterations

Linear constraints: $\begin{array}{cc} \min & f(x) \\ \text{s.t.} & Ax = b \end{array}$

(1) (a) Select feasible starting point for $Ax = b$;
$A = [B|C]$, B-basis, y-basic variable, u-independent variable.
   (b) Set reduced Hessian $H_R^0 = I$; set $k = 0$.

(2) Compute $f(x^k), \nabla f(x^k) = \begin{bmatrix} \nabla_y f \\ \nabla_u f \end{bmatrix}$.

(3) (a) Solve $B^T \lambda = -\nabla_y f$
   (b) Calculate reduced gradient, $g_R^T = \lambda^T C + \nabla_u f^T$.
   (c) Calculate steps Δu and Δy from $H_R^k \Delta u = -g_R$ and $B\Delta y = -C\Delta u$
   and set $\Delta x = \begin{bmatrix} \Delta y \\ \Delta u \end{bmatrix}$.

(4) Do line search for $x^{k+1} = x^k + \alpha\Delta x$, where $\alpha =$ step size to min $f(x)$ along Δu.

Figure 4.1 Newton correction to regain feasibility.

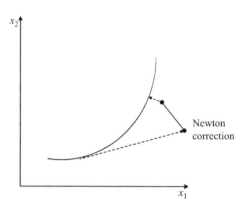

(5) Check convergence.

If $\left\|g_R^T g_R\right\|_2 \leq \varepsilon$, STOP. Else go to step (6).

(6) Update H_R^{k+1} with the BFGS formula, set $s^k = \Delta u$, $y^k = g_R\left(u^{k+1}\right) - g_R\left(u^k\right)$

Set $k = k + 1$, return to step (2).

(II) <u>Outer iterations for nonlinear constraints</u>

  (a) Linearize augmented Lagrangean (4.32).
  (b) Apply algorithm I for inner optimization.

We should note that this algorithm is also an infeasible path method in the sense that the optimization and convergence are carried out simultaneously. The code MINOS (Murtagh and Saunders, 2003) essentially implements this algorithm.

In order to increase robustness codes like CONOPT (Drud, 1994) and GRG (Lasdon et al., 1978) perform some Newton corrections to try to regain feasibility at each major iteration as shown in Fig. 4.1 making the code look more like a feasible path method.

4.3   Interior-Point Method

We reformulate the NLP in (4.1) with only equality constraints,

$$
\begin{aligned}
\min \quad & f(x) \\
\text{s.t.} \quad & c(x) = 0 \\
& x \geq 0,
\end{aligned}
\tag{4.33}
$$

which can be accomplished by replacing the inequalities by equalities and introducing slack variables, similarly as we did in (4.31).

We then apply Newton's method to the "barrier" problem (Fiacco and McCormick, 1968), in which the nonnegativity of the x_i variables is handled by introducing the penalty term $-\alpha \sum_i \ell n(x_i)$

Figure 4.2 Barrier function for the interior-point method.

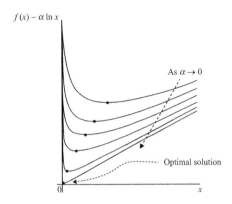

$$\min \quad f(x) - \alpha \sum_i \ell n(x_i)$$
$$\text{s.t.} \quad c(x) = 0 \tag{4.34}$$

which then gives rise to the barrier term in the objective function shown in Fig. 4.2 (Fiacco and McCormick, 1968). Notice that as $\alpha \to 0$ the barrier function approaches the objective function $f(x)$.

As shown in Marsten et al. (1990), (4.34) turns out to be equivalent to applying Newton's method to the KKT conditions with complementarity conditions, $\mu_i x_i = \alpha$, $\alpha > 0$. More specifically,

$$\nabla f(x) + A(x)\lambda - v = 0$$
$$Xv - \mu e = 0 \tag{4.35}$$
$$c(x) = 0,$$

where

$$e^T = [1, 1, 1, \ldots], \quad X = \text{diag}(x)$$
$$A = \nabla c(x), \quad W = \nabla_{xx} L(x, \lambda, v).$$

If we solve the systems of equations in (4.35) applying Newton's method, this leads to:

$$\begin{bmatrix} W & A & -I \\ A^T & 0 & 0 \\ V & 0 & X \end{bmatrix} \begin{bmatrix} d_x \\ d_\lambda \\ d_v \end{bmatrix} = - \begin{bmatrix} \nabla f + A\lambda - v \\ c \\ Xv - \mu e \end{bmatrix}. \tag{4.36}$$

The step size α is selected to ensure a sufficient decrease of a merit function using a "step to the boundary" rule with $\tau \sim 0.99$, that is, for $\alpha \in (0, \bar{\alpha}]$,

$$x_{k+1} = x_k + \alpha \, d_x$$
$$x_k + \bar{\alpha} \, d_x \geq (1 - \tau)x_k > 0$$
$$v_{k+1} = v_k + \bar{\alpha} \, d_v \geq (1 - \tau)v_k > 0 \tag{4.37}$$
$$\lambda_{k+1} = \lambda_k + \alpha \, (\lambda_+ - \lambda_k).$$

The above is essentially the basis of the codes IPOPT (Wächter and Biegler, 2006) and KNITRO (Byrd et al., 2006), both of which implement the interior-point method for solving NLP problems.

4.4   Comparison of NLP Algorithms

Based on the methods covered in this chapter, we briefly discuss the software available for these methods (see Appendix A), and their relative advantages and limitations.

SNOPT is a representative code implementing the successive-quadratic programming (SQP) algorithm. This method tends to be best for handling nonlinearities, and tends to require relatively few iterations. We should note that special implementations that are based on eliminating equations allow for solving large-scale problems (Biegler, 2010).

MINOS and CONOPT are representative codes for the reduced-gradient method. These codes tend to perform best for mostly linear constraints. Although they tend to require more iterations than the SQP method, the advantage is the implementation of the subproblems of the inner iterations that apply "LP Simplex methodology" given the linearity of the constraints. In fact, we should note that for the limiting case when problem (4.1) has linear objective function and constraints, which leads to an LP, MINOS reduces to the Simplex algorithm.

CONOPT is slower than MINOS but it is more robust as it tries to recover feasibility at each iteration.

IPOPT and KNITRO are representative codes of the interior-point method for NLP. These codes require relatively few iterations, but computational benefits are only realized with large-scale problems because in smaller problems the heavy overhead with the linear algebra at each iteration makes the performance of these methods slower.

4.5   Guidelines for Formulating NLP Models

To close this chapter, we briefly review some general guidelines for formulations of "good" NLP models as in the problem in (4.38) below, in which we assume that the objective and constraints are written in equation form,

$$
\begin{aligned}
\min \quad & f(x) \\
\text{s.t.} \quad & h(x) = 0 \\
& g(x) \leq 0 \\
& x \in R^n.
\end{aligned}
\tag{4.38}
$$

Some major guidelines are the following.

(1) It is important to try to specify lower, upper bounds on all variables

$$
x^L \leq x \leq x^U
\tag{4.39}
$$

as this leads to more constrained feasible regions, which in turn helps to obtain better starting points and make the solution of the NLP more robust.

(2) It is important to try to formulate whenever possible all the constraints as linear constraints, as this obviously helps all the methods, especially reduced-gradient methods. As a simple example, it is always better to reformulate the nonlinear inequality

$$\frac{x_1}{x_2} \leq 2 \tag{4.40}$$

as the linear equality

$$x_1 - 2x_2 \leq 0. \tag{4.41}$$

(3) Elimination of variables only pays off if the nonlinearities do not increase much in constraints, or even better if they can be transferred into the objective function.
   Consider as an example the NLP:

$$
\begin{aligned}
\min \quad & x_1 + 2x_2 \\
\text{s.t.} \quad & x_1 x_2 = 2 \\
& 0 \leq x_1 \leq 2 \\
& 0 \leq x_2 \leq 2.
\end{aligned}
\tag{4.42}
$$

If we eliminate from the equality constraint, $x_2 = 2/x_1$ then from the lower bound $x_2 \geq 2 \Rightarrow \frac{2}{x_1} \geq 0 \Rightarrow x_1 < \infty$, while from the upper bound $x_2 \leq 2 \Rightarrow \frac{2}{x_1} \leq 2 \Rightarrow x_1 \geq 1$. This then leads to the reformulated NLP,

$$
\begin{aligned}
\min \quad & x_1 + \frac{4}{x_1} \\
\text{s.t.} \quad & 1 \leq x_1 \leq 2,
\end{aligned}
\tag{4.43}
$$

which is smaller in size (reduced by one dimension), and with a convex nonlinearity only in the objective function. Therefore, (4.43) is a desirable formulation.
   Consider the second example:

$$
\begin{aligned}
\min \quad & x_1 + 2x_2 + 3x_3 \\
\text{s.t.} \quad & e^{x_1}(x_2 + x_3) = 10 \\
& 1 \leq x_1 \leq 2 \\
& 0.5 \leq x_2 \leq 2 \\
& 0.5 \leq x_3 \leq 2.
\end{aligned}
\tag{4.44}
$$

If we eliminate x_1, this leads to $x_1 = \ell n \left[\frac{10}{(x_2 + x_3)^2} \right]$. If we then substitute x_1, in (4.44) this leads to the reformulated NLP (4.45),

$$
\begin{aligned}
\min \quad & \ell n \left[\frac{10}{(x_2 + x_3)^2} \right] + 2x_2 + x_3 \\
\text{s.t.} \quad & e(x_2 + x_3)^2 \leq 10 \\
& - e^2(x_2 + x_3)^2 \leq -10 \\
& 0.5 \leq x_2 \leq 2 \\
& 0.5 \leq x_3 \leq 2,
\end{aligned}
\tag{4.45}
$$

which in fact introduces more nonlinearities than in (4.44) making the problem potentially more difficult to solve. This clearly shows that variable elimination is not always
advisable in NLP models.

(4) It is important to try to scale the variables and constraints. For instance, if we have the
bounded variables,

$$0 \leq x_1 \leq 1 \quad 0 \leq x_2 \leq 10\ 000,$$

we can define a new variable $z_2 = \dfrac{x_2}{10\ 000}$ so that the bounds become

$$0 \leq x_1 \leq 1 \quad 0 \leq z_2 \leq 1.$$

(5) The last guideline is given because Newton's method requires an initial guess within a
neighborhood of the solution. This means one should try to supply "good" initial
guesses, which of course is often not trivial!

EXERCISES

4.1 Given the problem (3.5) in Chapter 3:

$$\begin{aligned}
\min \quad & f(x) = (x_1 - 3)^2 + (x_2 - 4)^2 \\
\text{s.t.} \quad & 2x_1 + x_2 - 6 \leq 0 \\
& x_1 - x_2 - 4 \leq 0 \\
& x_1 \geq 1.8 \\
& x_2 \geq 0,
\end{aligned}$$

verify its solution with GAMS/MINOS, GAMS/CONOPT, GAMS/SNOPT,
GAMS/KNITRO, GAMS/IPOPT.

4.2 Assume we would like to solve the following NLP:

$$\begin{aligned}
\min \quad & \max_{j \in J} \left\{ f_j(x) \right\} \\
\text{s.t.} \quad & g(x) \leq 0 \\
& x \in R^n.
\end{aligned}$$

What difficulty will arise when trying to find a point that satisfies the Karush–Kuhn–
Tucker conditions in the above problem? Can you reformulate the problem to avoid this
difficulty?

4.3 (a) Show that for the linearly constrained NLP problem

$$\begin{aligned}
\min \quad & f(x) \\
\text{s.t.} \quad & Ax = b \quad (A\ m \times n \text{ matrix}, \quad m < n \text{ full rank}) \\
& x \in R^n
\end{aligned}$$

the quadratic program in the SQP algorithm reduces to

$$\min \quad \nabla f(x^i)^T d + \frac{1}{2} d^T \nabla_{xx} f(x^i) d$$

$$\text{s.t.} \quad Ax^i + Ad - b = 0.$$

(b) Assume that at a feasible point x^i the above quadratic program is solved. Determine if the predicted step above is identical to the one predicted by the reduced-gradient method.

4.4 Assume you apply the SQP algorithm to the quadratic programming problem:

$$\min \quad Z = c^T x + \frac{1}{2} x^T B x$$

$$\text{s.t.} \quad Ax \leq b$$

$$x \geq 0,$$

where B is a positive definite matrix. Determine under what assumptions the SQP algorithm will converge in one iteration and explain why.

4.5 Consider the design of a storage vessel that has the form of a cylinder. The required volume is 25 m³. The cost of the side of the cylinder is \$150/m², while the top and bottom cost \$190/m² and \$260/m², respectively. Formulate an NLP problem to determine the optimal dimensions of this vessel, and solve with GAMS/MINOS, GAMS/CONOPT and GAMS/SNOPT.

4.6 Consider the nonlinear programming problem:

$$\max \quad Z = f(x)$$

$$\text{s.t.} \quad h_j(x) \geq 0 \quad j = 1, \ldots, m$$

$$x^L \leq x \leq x^U.$$

Assuming that $f(x)$ and $h(x)$ are continuous and differentiable in the n-dimensional vector x, derive the Karush–Kuhn–Tucker conditions for the above problem.

4.7 Given are the two following optimization problems:

$$\text{P1}: \quad \min Z^1 = f(x)$$

$$\text{s.t.} \quad g(x) \leq 0$$

$$h(x) \leq 0$$

$$x \in R^n$$

$$\text{P2}: \quad \min Z^2 = f(x)$$

$$\text{s.t.} \quad g(x) + h(x) \leq 0$$

$$x \in R^n.$$

(a) Show that the optimal objective function values of the above problems obey the following inequality:

$$\left(Z^1\right)^* \geq \left(Z^2\right)^*.$$

(b) Does the above inequality rely on the assumption that the functions $g(x)$ and $h(x)$ are convex?

5 Linear Programming

5.1 Basic Theory

Linear programming (LP) models are among the most extensively used models in optimization (Saigal, 1995; Vanderbei, 2020). These are defined by the optimization of a linear objective function, subject to linear constraints and nonnegativity conditions on the continuous variables x. Specifically, LP models are given by the following formulation, which assumes minimization of the objective Z,

$$\begin{aligned} \min \quad & Z = c^T x \\ \text{s.t.} \quad & Ax = b \\ & x \geq 0, \end{aligned} \qquad \text{(LP)}$$

where x is an n-vector, A is an $m \times n$ matrix, c is the n-vector of cost coefficients, and the right-hand side b is an m-vector. Notice that $m < n$, which means there are nonzero degrees of freedom, assuming that the constraints are independent.

We should note the following about problem (LP).

(1) Inequality constraints can be transformed into equalities with slack variables. For instance $x_1 + x_2 \leq 2$ can be converted into an equality by adding the slack variable $x_3 \geq 0$, which leads to the equation $x_1 + x_2 + x_3 = 2$.

(2) Problem (LP) is a convex programming problem, and therefore it has a unique optimum solution.

(3) Problem (LP) is solvable in polynomial time (Karmarkar, 1984).

The two major methods for solving problem (LP) are: (a) the Simplex algorithm (Kantorovich, 1948; Dantzig, 1949, 1963) and (b) interior-point method (Karmarkar, 1984). Theoretically, the Simplex algorithm has exponential complexity, while the interior-point method has polynomial complexity. As we will see, the Simplex algorithm is closely related to the reduced-gradient method for NLP, while the interior-point method is also closely related to the interior-point methods for NLP. The major software packages for LP, CPLEX, XPRESS, and GUROBI implement both methods. Generally, the Simplex algorithm is computationally more efficient for solving problems with several thousand variables and constraints, while interior-point methods perform best in very large-scale problems.

In order to describe the Simplex algorithm, we first present some basic definitions and theoretical properties.

Figure 5.1 Optimal solution of an
LP lies at a vertex point of the
feasible region.

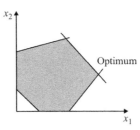

Definition 5.1 The feasible region $F = \{x | Ax = b, x \geq 0\}$ is defined by a polyhedron that is a closed convex region.

Since the objective function is linear and the feasible region is a polyhedron, the Simplex algorithm exploits the fact that the optimum lies at a vertex or extreme point, as seen in Fig. 5.1.

Definition 5.2 $x \in F$ is an extreme point if and only if for x such that,

$$x = \alpha x^1 + (1 - \alpha)x^2, \quad \alpha \in (0, 1), x^1, x^2 \in F, \quad \text{then } x^1 = x^2.$$

In other words, an extreme point cannot be obtained by a linear combination of different points $x^1 \neq x^2$ in F.

Consider partitioning the matrix of coefficients $A = [B | N]$, where the $m \times m$ basis matrix B is of full rank, and where N is the $m \times (n - m)$ matrix. Consider also partitioning the vector of variables x

$$x = \begin{bmatrix} x_B \\ x_N \end{bmatrix}, \tag{5.1}$$

where x_B is an m-vector of basic variables, and x_N is an $n - m$-vector of nonbasic variables. We can then express the linear constraints of problem (LP) as:

$$Ax = Bx_B + Nx_N = b. \tag{5.2}$$

If we set the nonbasic variables $x_N = 0$, this then implies that the basic variables can be computed as follows,

$$x_B = B^{-1}b, \tag{5.3}$$

where B^{-1} is assumed to exist as the basis B is assumed to be full rank.

As shown by the following theorem, the nonnegative basic variables computed from (5.3) and zero nonbasic variables define an extreme point in the feasible region of the LP.

Theorem 5.1 *A point x in (LP) is an extreme point of its corresponding feasible region F if and only if $x_B = B^{-1}b$, $x_B \geq 0$, $x_N = 0$.*

Proof Assume x is not an extreme point in F. This implies that x can be obtained as a linear combination of two distinct points x^1, x^2, that is,

$$x = ax^1 + (1 - a)x^2, \quad a \in (0, 1), x^1, x^2 \in F, \quad x^1 \neq x^2. \tag{5.4}$$

This implies for $x_N = 0 = ax_N^1 + (1 - a)x_N^2$. However, since $x_{N^1}, x_{N^2} \geq 0$ and $a \in (0, 1)$, this in turn implies $x_N^1 = x_N^2 = 0$, which contradicts the assumption that x_N is not an extreme point. Hence, $x_N = 0$ is an extreme point.

Consider now the basic variables, $x_B = B^{-1}b = ax_B^1 + (1 - a)x_B^2$. Since $x_B = B^{-1}b \geq 0$, is a unique solution and $a \in (0, 1)$, this implies $x_B = x_B^1 = x_B^2$, which contradicts the assumption that x_B is not an extreme point. Hence, $x_B = B^{-1}b, \; x_B \geq 0$ is an extreme point.   ∎

Based on Theorem 5.1, we can prove the following theorem.

Theorem 5.2 *The optimal solution x^* of (LP) lies at an extreme point of F.*

Proof Assume that x^* is not an extreme point. It then follows that x^* can be expressed as a linear combination of extreme points, that is, $x^* = \sum_i a_i x^i$, where x^i is an extreme point in F and $a_i \geq 0$, $\sum_i a_i = 1$. If we multiply x^* by c^T, we obtain

$$c^T x^* = \sum_i a_i c^T x^i. \tag{5.5}$$

But since by assumption x^* is the optimal solution, we have $c^T x^* < c^T x^i, \forall i$. If we multiply this inequality by a_i and take the corresponding sum over $a_i \geq 0$, this leads to $c^T x^* < \sum_i a_i c^T x^i$, which contradicts the assumption that x^* is not an extreme point. Hence, the optimal solution x^* of problem (LP) lies at an extreme point of F.   ∎

We should note that the optimal solution of problem (LP) is a global optimum because the objective function and the feasible region are convex. However, the global optimum may not be unique, as shown in Fig. 5.2 in which we present two alternate optima that have the same objective value.

The next important question that we need to address is to see how we can identify an extreme point in F as being an optimal solution. For this let us eliminate x_B from problem (LP) for a feasible basis; that is, from $Bx_B + Nx_N = b$, we obtain

$$x_B = -B^{-1}N, \quad x_N = B^{-1}b \geq 0. \tag{5.6}$$

If we substitute x_B in the objective function of the problem (LP) expressed in terms of basic and nonbasic variables, $Z = c_{B^T}x_B + c_{N^T}x_N$, we can define a reduced (LP′) in terms of the nonbasic variables x_N,

Figure 5.2 Alternate optima in an LP.

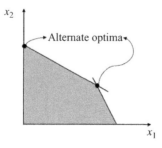

$$\min \quad Z = \left[c_B^T B^{-1} N - c_N^T\right] x_N + c_B^T B^{-1} b$$
$$\text{s.t.} \quad -x_N \le 0. \tag{LP'}$$

Comparing with Eq. (4.28) in Chapter 4 (the term $-\nabla_y f^T B^{-1} C + \nabla_u f^T$), the coefficient for x_N can be interpreted as the reduced gradient $g_{R^T} = -\left[c_B^T B^{-1} N - c_N^T\right]$. We can then establish the following theorem.

Theorem 5.3 *A necessary and sufficient condition for x^* to be an optimum of (LP') is given by the inequality $-c_B^T B^{-1} N - c_N^T \le 0^T$, which is equivalent to stating that $-g_{R^T} \le 0^T$.*

Proof From the Karush–Kuhn–Tucker conditions (3.31)–(3.33) (see Chapter 3), and defining ρ as the multiplier for the nonnegativity of x_N, a necessary and sufficient condition for (LP') is given by,

$$-c_B^T B^{-1} N + c_N^T - \rho^T = 0^T$$
$$\rho \ge 0, \quad -x_N \le 0, \quad \rho^T x_N = 0. \tag{5.7}$$

This implies $c_B^T B^{-1} N - c_N^T = -\rho^T$. But because $\rho \ge 0$ this implies $c_B^T B^{-1} N - c_N^T \le 0^T$, which proves the theorem. ∎

We should note that the multiplier ρ can be interpreted as the vector of reduced costs.

5.2 Simplex Algorithm

Given Theorems 5.1–5.3 in the previous section, we have all the elements required to derive the Simplex algorithm. Qualitatively what we need to do is to select the "correct" basis that defines the optimal extreme point. Since we cannot do this easily a-priori, we will develop an iterative "exchange" algorithm for generating alternative bases. The exchanges are based on the two following operations.

(a) Make one of the nonbasic variables a basic variable.
(b) Make one of the basic variables a nonbasic variable.

The generation of bases continues until the optimality condition given by Theorem 5.3 is satisfied, namely, $c_B^T B^{-1} N - c_N^T \le 0^T$, or equivalently until all the reduced costs ρ are positive. The specific criteria that are used for the exchanges for deciding which nonbasic variable should enter the basis and which basic variable should leave it are as follows.

(a) Which variable should enter the basis?

Assume $\bar{x}$ is a nonoptimal extreme point; then there exists at least one nonbasic variable x_{N_j} such that

$$c_B^T B^{-1} n_j - c_{N_j} > 0, \tag{5.8}$$

where n_j is the jth column of the submatrix N. Since there might be several nonbasic variables that satisfy (5.8), as a heuristic we select those with the largest positive left-hand

Figure 5.3 Selection of nonbasic variable with steepest descent.

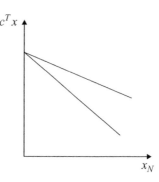

side (LHS) in (5.8) to enter the basis. Geometrically this can be interpreted as selecting the nonbasic variable with the steepest descent, as shown in Fig. 5.3.

(b) Which variable should leave the basis?

The main concern here is to ensure feasibility when making the choice on which variable should leave the basis. For this, let us consider the current basis, which can be represented as follows:

$$Bx_B + Nx_N = b. \tag{5.9}$$

Premultiplying by B^{-1} leads to

$$Ix_B + B^{-1}Nx_N = B^{-1}b, \tag{5.10}$$

which we rewrite as

$$Ix_B + Yx_N = \bar{b}, \tag{5.11}$$

where $Y = B^{-1}N$ and $\bar{b} = B^{-1}b$. From (5.10) for each basic variable x_{B_i} we have

$$x_{B_i} + \sum_k y_{ik}x_{Nk} = \bar{b}_i \quad i = 1, \ldots, m. \tag{5.12}$$

Assume the nonbasic variable x_{N_j} enters the basis. Since we want $x_{N_j} \geq 0$ and $x_{B_i} \geq 0 \quad \forall i$ to ensure feasibility, from (5.12) we obtain

$$x_{B_i} = \bar{b}_i - y_{ij}x_{N_j} \geq 0 \quad \forall i \tag{5.13}$$

from which it follows that

$$x_{N_j} \leq \bar{b}_i/y_{ij} \quad \forall i. \tag{5.14}$$

To ensure that (5.14) is satisfied for all i, we pick the rth basic variable with the smallest positive right-hand side (RHS), that is,

$$a_{rj} = \min_i \left\{ \bar{b}_i / y_{ij} \right\} \quad \text{for} \quad \bar{b}_i / y_{ij} > 0, \tag{5.15}$$

which is known as the min test ratio.

With the two exchange rules given by the largest LHS in (5.8) and by (5.15), we have the required criteria for generating a new basis when the current one is found not to be optimal, that is, not satisfying the conditions in Theorem 5.3.

To derive the Simplex algorithm from the objective function and constraints expressed as

$$Z - c_B^T x_B - c_N^T x_N = 0$$
$$B x_B + N x_N = b, \tag{5.16}$$

we represent (5.16) through the following Simplex tableau in which we explicitly display the objective Z, the basic and nonbasic variables x_B, x_N, and the RHS,

Z	x_B^T	x_N^T	RHS	
1	$-c_B^T$	$-c_N^T$	0	(objective)
0	B	N	b	(constraint rows).

(T)

In order to update the tableau we multiply the constraint rows by B^{-1}, multiplied by c_B^T, and we add them to the objective row, which is equivalent to a pivoting operation,

Z	x_B^T	x_N^T	RHS
1	0	$c_B^T B^{-1} N - c_N^T$	$c_B^T \bar{b}$
0	1	$B^{-1} N$	$\bar{b}$.

(5.17)

From the tableau (5.17), because the nonbasic variables x_N are zero at an extreme point, we can directly read the value of the basic variables $x_B = B^{-1}b$ as in (5.3). Furthermore, the coefficients of the nonbasic variables in the objective function $c_B^T B^{-1} N - c_N^T$ correspond to the negative of the reduced gradient $-g_{R^T}$. From Theorem 5.3 we can then check if $c_B^T B^{-1} N - c_N^T \leq 0^T$. If that is the case, we have found the optimal solution of the LP with the current basis; otherwise, we know that the current basis is not optimal, and we need to find a new one using the exchange rules described above. In summary, starting from the tableau (T) we perform successive updates through pivoting operations until the optimality conditions in (5.17) are satisfied.

To complete the description of the Simplex algorithm, we need to determine how we select the initial basis. The simplest case is when the constraints of the LP are only inequalities, $A'x \leq b$, with the RHS $b \geq 0$. In this case, by introducing slack variables $s \geq 0$, we can rearrange the LP as follows:

$$\begin{aligned} \min \quad & Z = c^T x \\ \text{s.t.} \quad & A'x + Is = b \\ & x \geq 0 \quad s \geq 0. \end{aligned} \tag{5.18}$$

Because $x = 0$ is a feasible point, the initial basis can simply be chosen as the identity matrix corresponding to the slack variables s, which are the initial basic variables, while the x variables are the initial nonbasic variables.

Because (5.18) is a special case, we need a more general procedure for problem (LP). We assume that the constraints are rearranged as follows: $A''x = b, b \geq 0$. We then introduce nonnegative artificial variables q and reformulate the LP as follows:

$$\min \quad Z = c^T x + Mq$$
$$\text{s.t.} \quad A''x + Iq = b \qquad\qquad (5.19)$$
$$x \geq 0 \quad q \geq 0,$$

where $M > 0$ is a large parameter. The initial basis can then be selected by choosing the variables q to be basic variables with the corresponding identity matrix for the basis. An alternative approach is to use a two-phase procedure where in phase I we solve for $Z = \min e^T q$ (i.e., feasibility problem) until we find a feasible point $(q = 0)$; next in phase II we optimize the original objective $Z = c^T x$.

In summary, the basic steps of the Simplex algorithm for solving the problem (LP) are as follows.

Step 1 Select initial basis $B = I$, set $N = A'$ or A'' (from (5.18) or (5.19)), and partition:

$$x = \begin{bmatrix} x_B \\ x_N \end{bmatrix}, c = \begin{bmatrix} c_B \\ c_N \end{bmatrix}.$$

Step 2 Define Simplex tableau.

Z	x_B^T	x_N^T		RHS	
1	$-c_B$	$c_B^T B^{-1} N - c_N^{T(1)}$	$c_B^T b^{(1)}$	Objective row
	I	$B^{-1} N^{(2)}$	b	Constraint rows

(1) At first iteration $c_B = 0$ for (5.18); (2) matrix N has columns n_k and elements n_{ij}.

Step 3

(a) If $c_B^T B^{-1} n_k - c_{N_k} \leq 0$ for all nonbasic variables k STOP. Otherwise select variable x_{N_j} with largest $c_B^T B^{-1} n_j - c_{N_j} > 0$ to enter basis.
(b) To decide which basic variable leaves basis, compute b_i / n_{ij}, $i = 1, \ldots, m$ and select row r with smallest positive value (min test ratio).

Step 4

(a) Select element n and pivot on matrix $\begin{bmatrix} I & B^{-1}N & b \end{bmatrix}$ to make their coefficient 1 and with zero columns for n_{ij} $i \neq r$, $i = 1, \ldots, m$.
(b) Add to objective row the new rth row multiplied by coefficient c_j (cost coefficient of variable j entering basis).
(c) Return to step 2 and continue until step 3a is satisfied.

5.3   Numerical Example

Consider the following LP:

$$\min \quad Z = -x_1 - 3x_2$$
$$\text{s.t.} \quad x_1 + x_2 \leq 4$$
$$x_1 + 2x_2 \leq 6 \tag{5.20}$$
$$x_1, x_2 \geq 0,$$

which has the simpler structure of only inequalities and a positive RHS (see (5.18)). To convert the inequalities into equalities we introduce the slack variables, x_3, x_4, which then yield the LP

$$\min \quad Z = -x_1 - 3x_2$$
$$\text{s.t.} \quad x_1 + x_2 + x_3 = 4$$
$$x_1 + 2x_2 + x_4 = 6 \tag{5.21}$$
$$x_1, x_2 \geq 0 \quad x_3, x_4 \geq 0.$$

The feasible region of the LP in (5.20) is displayed in Fig. 5.4. Note that the slack variables x_3, x_4 indicate the deviation from the boundary of the two corresponding inequalities.

The tableau at iteration 1 is shown below.

		Z	x_1	x_2	x_3	x_4	RHS
R1	*basis*	1	1	3	0	0	0
R2	x_3		1	1	1	0	4
R3	x_4		1	2	0	1	6

From this tableau we can infer that the objective $Z = 0$, the basic variables $x_3 = 4$, $x_4 = 6$, and the nonbasic variables $x_1 = x_2 = 0$. Because the coefficients of these variables are positive $[1, 3] > [0, 0]$ the corresponding extreme point is nonoptimal. Because 3 is the larger value, we select x_2 as the variable that enters the basis. We then perform the min test ratio. For R2 we obtain $4/1 = 4$, while for R3 we obtain $6/2 = 3$. As the latter is smaller, we select x_4 as the variable that leaves the basis.

Figure 5.4 Feasible region for example problem.

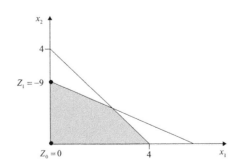

We next perform the pivoting operations shown in the left of the table below where they are applied in the order R3/2, R2 − R3/2, R1 − 3R3/2.

		Z	x_1	x_2	x_3	x_4	RHS
R1 − 3R3/2	basis	1	$-\frac{1}{2}$	0	0	$-\frac{3}{2}$	−9
R2 − R3/2	x_3		$\frac{1}{2}$	0	1	$-\frac{1}{2}$	1
R3/2	x_2		$\frac{1}{2}$	1	0	$\frac{1}{2}$	3

Note that the basis is defined by x_2, x_3 with its corresponding identity matrix for the basis. From this table we can then determine that the basic variables $x_2 = 3$, $x_3 = 1$ and $Z = -9$. Since the coefficients of $c_{B^T}B^{-1}N - c_{N^T} = [-\frac{1}{2}, -\frac{3}{2}] < [0, 0]$, we can assert that the extreme point $x_1 = 0$, $x_2 = 3$, $x_3 = 1$, $x_4 = 0$, corresponds to the optimal solution of the LP in (5.20), and hence we can stop the search.

EXERCISES

5.1 Given is the linear programming problem

$$\min \quad Z = c^T x$$
$$\text{s.t.} \quad Ax = b$$
$$x \geq 0,$$

where A is an $m \times n$ matrix with rank m. Show that, for a feasible selection of the basis B, the Lagrange multipliers λ of the equations are given by

$$\lambda = -\left(B^T\right)^{-1} c_B,$$

where c_B is the cost coefficient of the basic variables.

5.2 Given is the following linear programming problem:

$$\min \quad Z = -2x_1 - 3x_2$$
$$\text{s.t.} \quad x_1 + x_2 \leq 4$$
$$x_1 + 2x_2 \leq 6$$
$$x_1 \geq 0, \quad x_2 \geq 0.$$

(a) Plot the contours of the objective and the feasible region. Determine the optimum by inspection.
(b) Solve the problem using the Simplex algorithm.

6   Mixed-Integer Programming Models

The most general mixed-integer problems, a generalization of nonlinear programs (NLP), are given by mixed-integer nonlinear programs (MINLP) (Lee and Leyffer, 2012; Trespalacios and Grossmann, 2014) involving continuous and integer variables, x, y,

$$
\begin{aligned}
\min \quad & Z = f(x, y) \\
\text{s.t.} \quad & h(x, y) = 0 \\
& g(x, y) \leq 0 \\
& x \in R^n \quad y \in \{0, 1\}^m.
\end{aligned}
\quad \text{(MINLP)}
$$

The most general MINLP has in fact the y variables being specified as integer variables $y \in Z_+^m$. However, we restrict ourselves to 0-1 variables for two reasons. First, the 0-1 variables tend to be the most common ones arising in applications. Second, an integer variable or even an arbitrary discrete variable can always be expressed in terms of 0-1 variables.

A case of special interest is the mixed-integer linear program (MILP) (Nemhauser and Wolsey, 1988; Wolsey, 1998) given by linear objective function and constraints,

$$
\begin{aligned}
\min \quad & Z = a^T x + b^T y \\
\text{s.t.} \quad & Ax + By \leq d \\
& x \geq 0 \quad y \in \{0, 1\}^m.
\end{aligned}
\quad \text{(MILP)}
$$

If there are no y variables, the MILP reduces to an LP. If there are no x variables, the MILP reduces to a pure integer program IP.

6.1   Modeling with 0-1 Variables

6.1.1   Motivating Examples

In this section we first introduce three motivating examples of constraints in terms of 0-1 variables to emphasize the point that while the constraints can intuitively be expressed as nonlinear constraints, they can also be written as linear constraints (Williams, 1999).

Let

$$
y_j = \begin{cases} 1 \text{ if } j \text{ true} \\ 0 \text{ if } j \text{ false}. \end{cases}
\tag{6.1}
$$

Figure 6.1 Relaxations of (a) constraint (6.2) and (b) constraint (6.3).

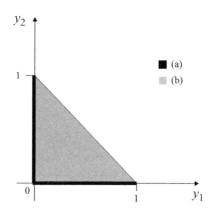

First, consider the case where at most one y_j is true (i.e., $y_j = 1$), for simplicity say y_1, y_2. If we express this constraint in nonlinear form it is given by (6.2),

$$y_1 \cdot y_2 = 0. \tag{6.2}$$

However, we can also express this constraint with a linear inequality

$$y_1 + y_2 \leq 1. \tag{6.3}$$

As shown in Fig. 6.1, the relaxation of constraint in (6.2) with $y_1 \geq 0$, $y_2 \geq 0$ defines a nonconvex feasible region (a), while the relaxation of constraint (6.3) with $y_1 \geq 0$, $y_2 \geq 0$ gives rise to a convex feasible region (b). Therefore, although constraint (6.2) is valid in the sense that it represents the condition stated, its mathematical properties are less desirable than the linear constraint in (6.3).

As a second case, consider the variable $y = 0, 1$. A valid nonlinear constraint to represent this binary variable as continuous, is the nonlinear inequality with the 0-1 bounds on y,

$$y(1 - y) \leq 0 \quad 0 \leq y \leq 1. \tag{6.4}$$

While (6.4) avoids the use of 0-1 variables, the drawback is that it is nonconvex. This can be seen by rearranging (6.4) as

$$y - y^2 \leq 0, \tag{6.5}$$

where it is clear that $-y^2$ is a concave function.

Finally, as a third case, assume we want to represent the condition if $y = 1 \Rightarrow h(x) = 0$. A simple nonlinear constraint representing this condition, which is nonconvex, is given by

$$yh(x) = 0, \tag{6.6}$$

where it can clearly be seen that $h(x) = 0$ if $y = 1$, while the constraint holds for $y = 0$ for any $h(x)$.

One can also reformulate the constraint in (6.6) through the "big-M" inequalities which are linear in the binary variable y, where M is a sufficiently large parameter,

$$-M(1-y) \leq h(x) \leq M(1-y). \tag{6.7}$$

Notice that if $y = 1$, (6.7) reduces to $h(x) = 0$, because $0 \leq h(x) \leq 0$, while for $y = 0$ (6.7) reduces to the inequality $-M \leq h(x) \leq M$, which is redundant for sufficiently large M. In practice, however, one should not select a very large value of M as it adversely affects the continuous relaxation of the mixed-integer programming problem.

The three examples, then, show that while nonlinear inequalities may represent 0-1 conditions, they have the undesirable property of being nonconvex, which would necessitate global-optimization methods (see Chapter 12). The main message from these examples is that one should always try to model 0-1 variables y so that they appear linearly in the mixed-integer programming model. Otherwise, expressing them in nonlinear form usually leads to nonconvexities.

6.1.2   Modeling with Linear 0-1 Variables y_j

In this section we present several major classes of constraints and functions that can be expressed with linear 0-1 variables (Nemhauser and Wolsey, 1988; Williams, 1999).

(I) Multiple-choice constraints
  As implied by the term characterizing these constraints, these deal with choices of values for the binary variables.
  (a) Select at least one

$$\sum_{j \in J} y_j \geq 1. \tag{6.8}$$

  (b) Select exactly one

$$\sum_{j \in J} y_j = 1. \tag{6.9}$$

  (c) Select not more than one

$$\sum_{j \in J} y_j \leq 1. \tag{6.10}$$

  As we will see later (Section 7.1 in Chapter 7), (a) can be interpreted as an inclusive OR, (b) as an exclusive OR, and (c) is often denoted as a knapsack constraint. We should also note that the logic operator AND defines imposing several constraints. For instance, consider: (select 1 or 2) AND (select 3 or 4). This logic statement can be translated as the two following inequalities (see Section 7.1 for more details):

$$\begin{aligned} y_1 + y_2 &\geq 1 \\ y_3 + y_4 &\geq 1. \end{aligned} \tag{6.11}$$

(II) If then condition

Assume we state the condition "If select y_k then select y_j" $\left(y_k \Rightarrow y_j \right)$, this condition can be written as the linear inequality

$$y_k - y_j \leq 0. \tag{6.12}$$

This can be easily verified as follows:

if $y_k = 1 \Rightarrow y_j = 1$; if $y_k = 0 \Rightarrow y_j = 0$ or 1.

We should note that the condition iff (if and only if) can be expressed as the equality

$$y_k = y_j. \tag{6.13}$$

(III) Disjunction

Consider the disjunctive constraint $f_1(x) \leq 0$ or $f_2(x) \leq 0$, meaning either f_1 or f_2 must be satisfied (i.e., less than or equal to zero). This disjunction can be expressed with the two following constraints that are linear in y and where M is a sufficiently large parameter:

$$\begin{aligned} f_1(x) &\leq M(1 - y_1) \\ f_2(x) &\leq M(1 - y_2) \\ y_1 + y_2 &= 1. \end{aligned} \tag{6.14}$$

We can easily verify from (6.14) that if $y_1 = 1$ then $y_2 = 0$ and we have $f_1(x) \leq 0$, $f_2(x) \leq M$, while if we have $y_1 = 0$ then $y_2 = 1$ and we have $f_2(x) \leq M$, $f_2(x) \leq 0$.

Later in Section 7.2 in Chapter 7 we will study in more detail the topic of disjunctive programming where more general forms of disjunctions are considered.

(IV) Discontinuous domains and functions

A prime example for this case is the fixed-charge cost model, which is widely used in optimization planning models

$$C = \begin{cases} \alpha + \beta x & L \leq x \leq U \\ 0 & x = 0, \end{cases} \tag{6.15}$$

and which is displayed in Fig. 6.2. The basic idea is that we have a discontinuous domain given by $x = 0$ and $L \leq x \leq U$. Furthermore, when $x = 0$ the cost $C = 0$, while for $L \leq x \leq U$ the cost is defined by the linear function $\alpha + \beta x$, where α represents the fixed cost while β is the variable-cost coefficient.

Figure 6.2 Fixed-cost function.

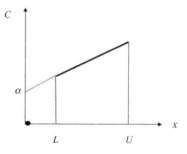

The fixed-cost function in (6.15) can be represented with linear 0-1 variables by

$$C = \alpha y + \beta x$$
$$Ly \leq x \leq Uy \qquad\qquad (6.16)$$
$$x \geq 0, y = 0, 1.$$

We can verify that if $y = 0$ then $x = 0$ and $C = 0$, while if $y = 1$ then $L \leq x \leq U$ and $C = \alpha + \beta x$. We should note that usually $L = 0$ with which the domain of x becomes continuous.

6.1.3 Some Common IP Problems

In this section we present some classic integer programming models to again illustrate the use of linear 0-1 variables (Nemhauser and Wolsey, 1988; Conforti et al., 2014).

6.1.3.1 Assignment Problem

This problem can be stated as follows. Given is a set of n jobs i and given is a set of n machines j (see Fig. 6.3). Also given is the cost c_{ij} for assigning job i to machine j. The problem then consists of finding the minimum cost assignment of the jobs to the machines.

To formulate the corresponding optimization problem we introduce the binary variables y_{ij} such that $y_{ij} = 1$ if job i is assigned to machine j at cost c_{ij}, and $y_{ij} = 0$ otherwise. The assignment problem can then be formulated as the following integer programming problem,

$$\min \quad Z = \sum_{i=1}^{n}\sum_{j=1}^{n} c_{ij} y_{ij}$$

$$\text{s.t.} \quad \sum_{j=1}^{n} y_{ij} = 1 \quad i = 1, \ldots, n$$

$$\sum_{i=1}^{n} y_{ij} = 1 \quad j = 1, \ldots, n \qquad \text{(AP)}$$

$$y_{ij} = 0, 1.$$

In (AP) we minimize the assignment cost subject to the constraints that for every job i only one machine j can be assigned, and for every machine j only one job i can be assigned. These assignment constraints represent exclusive OR statements.

An interesting property of problem (AP) is that it can be solved as an LP with continuous variables y_{ij}, such that $0 \leq y_{ij} \leq 1$. The reason is that all the extreme points of the polytope

Figure 6.3 Assignment of n jobs to n machines.

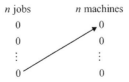

Figure 6.4 Plant location problem.

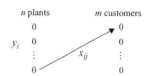

defined by assignment constraints correspond to 0-1 values for the continuous variables y_i. In general, IPs that have this property are said to have a totally unimodular matrix (Nemhauser and Wolsey, 1988).

6.1.3.2 Plant Location Problem

Consider n potential plants that are to serve m customers j with demands d_j (see Fig. 6.4). Each plant i has a fixed cost f_i. Also, the combined transportation and manufacturing cost from a plant i to customer j is given by the cost c_{ij}. The problem then consists of selecting the plants that serve the customers so as to satisfy the demands at minimum total cost.

If we denote by y_i the 0-1 variables for selecting each plant i, and by x_{ij} the amount of material sent from plant i to customer j, the MILP problem can be formulated as follows:

$$\min \quad Z = \sum_{i=1}^{n} f_i y_i + \sum_{i=1}^{n}\sum_{j=1}^{m} c_{ij} x_{ij}$$

$$\text{s.t.} \quad \sum_{i=1}^{n} x_{ij} = d_j \qquad j = 1, \ldots, m \tag{PL}$$

$$\sum_{j=1}^{n} x_{ij} - U_i y_i \leq 0 \quad i = 1, \ldots, n$$

$$y_i = 0, 1 \quad x_{ij} \geq 0,$$

where U_i represents the maximum capacity of plant i. In problem (PL) the objective function includes the fixed costs of the selected plants and their corresponding manufacturing/transportation costs. The first equality states that for each customer j the amounts x_{ij} sent by all the plants i have to satisfy the corresponding demands d_j. The second inequality states that the amount of material sent from each plant i to all the markets cannot exceed the maximum capacity U_i. In contrast to the assignment problem (AP), problem (PL) is known to be NP-hard (Cornuéjols et al., 1990), meaning that in the worst case the computational time scales exponentially with the problem size of the MILP.

6.1.3.3 Knapsack Problem

The knapsack is another classical combinatorial problem. It assumes that we are given n objects ($i \in I$) of given weight w_i and price p_i. We are also given a knapsack that can hold a maximum weight W of objects (see Fig. 6.5). The problem then consists of selecting those objects that can be accommodated in the knapsack and that maximize their value.

Figure 6.5 Knapsack problem.

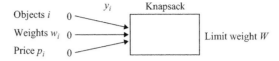

Figure 6.6 Set-covering problem for searching m units information in n files.

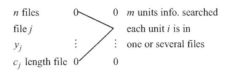

This problem can be formulated as the pure 0-1 integer programming problem (KP) by defining the binary variable y_i , where $y_i = 1$ if object i is selected and $y_i = 0$ otherwise:

$$\max \quad Z = \sum_{i=1}^{n} p_i y_i$$

$$\text{s.t.} \quad \sum_{i=1}^{n} w_i y_i \leq W \qquad \text{(KP)}$$

$$y_i = 0, 1 \quad i = 1, \ldots, n.$$

In (KP) the objective corresponds to maximizing the value of the selected objects, while the inequality limits the weights of the selected objects not to exceed the maximum weight W. Despite its apparent simplicity, (KP) is NP-hard (Nemhauser and Wolsey, 1988).

6.1.3.4 Set-Covering Problem

The set covering is another classical combinatorial problem. We will state it here in terms of a computer search problem (see Fig. 6.6) (Nemhauser and Wolsey, 1988). Assume that we are given n computer files containing units of information i, which are represented by the set that indicates the location of the unit of information i in file j, $P_i = \{i | \text{unit } i \text{ is in file } j\}$. Furthermore, assume that the size of each file j is given by c_j (e.g., in megabytes). The problem then consists of deciding which files to open for reading so as to retrieve the required units of information $i \in P_j$ and so as to minimize the total size of the files that are searched.

If we define the binary variable $y_j = 1$ if file j is searched, the set-covering problem (SC) can be formulated as follows:

$$\min \quad Z = \sum_{j=1}^{n} c_j y_j$$

$$\text{s.t.} \quad \sum_{j=1}^{n} a_{ij} y_j \geq 1 \quad i = 1, \ldots, m \qquad \text{(SC)}$$

$$y_j = 0, 1,$$

in which we minimize in the objective function the total length of the files, subject to the constraints that the m units of information be retrieved. In order to represent such a constraint we define the coefficients $a_{ij} = \begin{cases} 1 & \text{if } i \in P_j \\ 0 & \text{otherwise} \end{cases}$ so that each information unit i is accessed through a file j. As an example, consider that unit of information 1 is present in files 1 and 2 but not in file 3. This then means that we set the coefficients as $a_{11} = 1$, $a_{12} = 1$, and $a_{13} = 0$.

6.1.3.5 Traveling Salesman Problem

The traveling salesman problem (TSP) is perhaps the most extensively studied integer programming problem (Laporte, 1992). The TSP considers n cities for which the distance or cost c_{ij} is given for every pair of cities i, j $(i \neq j)$. The problem then consists of finding the shortest or cheapest tour so that every city is visited only once. This condition is equivalent to finding a Hamiltonian path in the network defined by the n cities and their corresponding pairwise arcs (see Fig. 6.7).

In order to formulate this problem, we define the binary variables y_{ij} such that:

$$y_{ij} = \begin{cases} 1 & \text{if go from city } i \text{ to city } j \\ 0 & \text{otherwise.} \end{cases}$$

The TSP can then be formulated as follows:

$$\min \quad Z = \sum_{i=1}^{n}\sum_{j=1}^{n} c_{ij} y_{ij}$$

$$\text{s.t.} \quad \sum_{j=1}^{n} y_{ij} = 1 \quad i = 1, \ldots, n$$

$$\sum_{i=1}^{n} y_{ij} = 1 \quad j = 1, \ldots, n \qquad \text{(TSP)}$$

$$\sum_{i \in Q}\sum_{j \in \bar{Q}} y_{ij} \geq 1 \quad \forall Q \subseteq V, \quad Q \neq \emptyset$$

$$y_{ij} \geq 0 \quad i = 1, \ldots, n, \quad j = 1, \ldots, n.$$

Figure 6.7 Traveling salesman problem network for four cities A, B, C, and D.

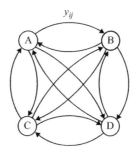

Figure 6.8 Two subtours for the six-city problem.

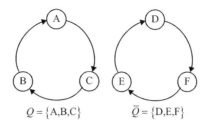

$Q = \{A,B,C\}$ $\bar{Q} = \{D,E,F\}$

The first two constraints correspond to assignment constraints. The first constraint states that for every city i there must be exactly one outgoing link to city j. The second constraint states that for every city j there must be exactly one incoming link from city i. These constraints, however, are not sufficient to guarantee a closed Hamiltonian path. For instance, for the case of six cities A, B, C, D, E, and F, one might obtain the two subtours shown in Fig. 6.8, which satisfy the assignment constraints. To make these infeasible, it is necessary to add the third set of constraints, which are known as the subtour elimination constraints. These constraints ensure that for every subset Q of the set of cities V (or vertices) there is at least one link with its complement $\bar{Q}$. The difficulty with this set of constraints is that there is an exponential number of them, which makes the TSP to be NP-hard.

We should also note that to avoid links from city i to the same city i (self loops) one can specify an infinite cost or distance, that is $c_{ij} = \infty$. Finally, when $c_{ij} = c_{ji}$ the TSP problem is symmetric, while if $c_{ij} \neq c_{ji}$ the TSP is asymmetric. Generally, the symmetric TSP is more difficult to solve as it will tend to give rise to a larger number of subtours. For more on the TSP see Laporte (1992), Cook (2012).

EXERCISES

6.1 Assume that in an MILP model you have an integer variable $n = 1, 2, \ldots, 8$, and that you want to express it in terms of 0-1 variables y_i. Determine the constraints that are needed to express n as:
   (a) a linear combination of the binary variables; and
   (b) a binary expansion in terms of binary variables. Discuss the relative advantages and disadvantages of both formulations.
6.2 The following inequality for 0-1 variables $y = 0, 1$, $i = 1, \ldots, n$, in which we partition these variables into the subsets, B, N, is known as an integer cut,

$$\sum_{i \in B} y_i - \sum_{i \in N} y_i \leq |B| - 1,$$

because its purpose is to make infeasible the choice of the following values of the binary variables, $y_i = 1$, $i \in B$, $y_i = 0$, $i \in N$, and given that $|B|$ is the cardinality of the set B. Explain why the constraint for the integer cut is correct.

7 Systematic Modeling of Constraints with Logic

The major goal of this chapter is to show that constraints involving 0-1 variables or general 0-1 mixed-integer constraints can be modeled systematically by applying concepts of propositional logic and disjunctive programming. This is in contrast with the models presented in Chapter 6 that were largely formulated by inspection or intuition.

7.1 Modeling 0-1 Constraints with Propositional Logic

In this section we will show that linear 0-1 constraints can be systematically derived using logic expressions for the constraints in the form of propositional logic (Raman and Grossmann, 1994; Williams, 1999; Hooker, 2002). In particular, propositional logic deals with symbolic expressions that involve the operators: OR (inclusive OR), AND, NOT (negation), and $\Rightarrow$ (implication). The operators exclusive OR and $\Leftrightarrow$ (if and only if) are treated as special cases.

To express a logic proposition, let p_j be a literal representing a selection or action. We can associate to that literal a binary variable y_j such that $y_j = 1$ means that the literal is true, while $y_j = 0$ means that it is false. Applying the basic operators to the literals p_j can be represented in terms of the binary variables as follows:

NOT: $\neg p_j$ by $1 - y_j$

OR: $p_i \vee p_j$ by $y_i + y_j \geq 1$

AND: $p_i \wedge p_j$ by $y_i = 1, y_j = 1$

$\Rightarrow$: $p_i \Rightarrow p_j \Leftrightarrow \neg p_i \vee p_j$ by $1 - y_i + y_j \geq 1$ or $y_i \geq y_j$.

Note that the implication $p_i \Rightarrow p_j$ makes use of the equivalent contraposition $\neg p_i \vee p_j$ to derive the corresponding inequality $y_i \geq y_j$.

The special cases of exclusive OR (XOR) and equivalence ($\Leftrightarrow$) are represented as follows:

XOR $p_i \underline{\vee} p_j$ by $y_i + y_j = 1$

EQUIVALENCE $p_i \Leftrightarrow p_j$ by $y_i = y_j$.

We should also note that $p_i \vee p_j$, which involves literals separated by inclusive OR operators, is known as a logic clause. A clause is a basic building block for the conjunctive normal form (CNF), which is given by a conjunction of clauses, as will be discussed below. In order to convert logic expressions into the CNF, which is a convenient form that can be readily transformed into linear 0-1 inequalities, we will make use of the two following properties from Boolean algebra.

(1) De Morgan's Theorem

    (a) $\neg(A \vee B) \Leftrightarrow \neg A \wedge \neg B$
    (b) $\neg(A \wedge B) \Leftrightarrow \neg A \vee \neg B$

(2) Distribution of OR

$(A \wedge B) \vee C \Leftrightarrow (A \vee C) \wedge (B \vee C)$.

These equivalences can be easily proved by applying Truth Tables from logic (Williams, 1999).

As we can see, De Morgan's Theorem applies the negation operator $\neg$ inwards of expressions containing the OR ($\vee$) or AND ($\wedge$) operators, while the distribution applies the OR operators on a conjunction of A and B.

The CNF is formally defined as a logic expression composed by the conjunction (AND) of clauses, which are composed of literals separated by the OR operator. That is, the CNF can be expressed as follows:

$$\Omega = \bigwedge_{j \in J} \left[\bigvee_{i \in I_j} p_i \right]. \tag{7.1}$$

An example is the logic expression, $(p_1 \vee \neg p_2) \wedge (\neg p_1 \vee p_3)$, which consists of two clauses. Notice that the literals may or may not be negated. From the above representation for a clause in terms of 0-1 variables, it readily follows that the two clauses can be represented by the two following inequalities

$$y_1 + 1 - y_2 \geq 1, \quad 1 - y_1 + y_3 \geq 1, \tag{7.2}$$

which can be rearranged as follows:

$$y_1 - y_2 \geq 0, \quad -y_1 + y_3 \geq 0. \tag{7.3}$$

Therefore, it is clear that if the logic is given as a CNF, one can readily transform it into linear 0-1 inequalities. The problem is that arbitrary logic is usually not expressed as a CNF. However, it turns out that one can systematically convert logic expressions in propositional logic by using a recursive procedure developed by Clocksin and Mellish (1994), and which was used in the logic program Prolog. The procedure to convert propositional logic into a CNF consists of applying the following steps in a recursive fashion until the CNF is reached.

(1) Remove implication.
(2) Move negation inwards by applying De Morgan's Theorem.
(3) Recursively distribute OR over the AND operator.

Once the CNF is obtained it is easy to translate it into inequalities as shown in the examples below.

7.1.1 Example 1 of Logic Proposition

Consider the logic statement "If select flash, then select distillation and not membrane" in a process-synthesis problem. The logic can be expressed in symbolic form as follows:

$$p_F \Rightarrow \vee(p_D \wedge \neg p_M). \tag{7.4}$$

We apply the three steps indicated above.

(1) Remove the implication leads to $\neg p_F \vee (p_D \wedge \neg p_M)$.
(2) There is no need to apply De Morgan's Theorem, so we proceed to distribute the OR over the AND, which yields, $(\neg p_F \vee p_D) \wedge (\neg p_F \vee \neg p_M)$. This corresponds to a conjunction of two clauses, that is it is a CNF, which in turn can then be translated into the inequalities

$$1 - y_F + y_D \geq 1, \quad 1 - y_F + 1 - y_M \geq 1, \tag{7.5}$$

which can be rearranged as follows:

$$y_D \geq y_F, \quad y_F + y_M \leq 1. \tag{7.6}$$

Notice that when $y_F = 1(\text{true}) \Rightarrow y_D = 1(\text{true})$, and $y_M = 0(\text{false})$. On the other hand, when $y_F = 0$ it implies $y_D = 0, 1$, and $y_M = 0, 1$.

7.1.2 Example 2 of Logic Proposition

Consider the following logic statement:

$$(p_1 \wedge p_2) \Rightarrow (p_3 \wedge p_4). \tag{7.7}$$

Because it corresponds to an implication, $A \Rightarrow B$, it is equivalent to $\neg A \vee B$. Hence, substituting, we have from De Morgan's Theorem and the distribution,

$$\neg A = \neg(p_1 \wedge p_2) \Leftrightarrow \neg p_1 \vee \neg p_2 = C$$
$$\neg A \vee B = C \vee B = C \vee (p_3 \wedge p_4) \Rightarrow (C \vee p_3) \wedge (C \vee p_4).$$

And, finally, we obtain

$$(\neg p_1 \vee \neg p_2 \vee p_3) \wedge (\neg p_1 \vee \neg p_2 \vee p_4), \tag{7.8}$$

which is in CNF. Rewriting both clauses in terms of 0-1 variables yields the inequalities

$$1 - y_1 + 1 - y_2 + y_3 \geq 1$$
$$1 - y_1 + 1 - y_2 + y_4 \geq 1, \tag{7.9}$$

which can be rearranged as

$$y_3 \geq y_1 + y_2 - 1$$
$$y_4 \geq y_1 + y_2 - 1. \tag{7.10}$$

Notice that when $y_1 = 1$, $y_2 = 1$, the inequalities in (7.10) yield $y_3 = 1$, $y_4 = 1$, which verifies the truth of the proposition in (7.7).

7.2 Modeling of Disjunctions

In this section we consider linear disjunctions of the following form (Balas, 1979; Balas, 2018):

$$\bigvee_{j \in D} \left[A_j x \le b_j \right], \tag{7.11}$$

where $j \in D$ is the set of disjuncts or disjunctive terms, x is an n-vector of continuous variables, A is an $m \times n$ matrix, and b_j is an m-vector of right-hand sides.

An example of a disjunction is the following condition that arises in batch scheduling problems in which it is stated that either A is manufactured before B, or B is manufactured before A. If the start times of A and B are given by t_A and t_B, respectively, and the processing times of A and B are given by p_A and p_B, respectively, the logic condition can be expressed as follows:

$$(t_A + p_A \le t_B) \vee (t_B + p_B \le t_A). \tag{7.12}$$

The disjunct on the left states that the start time of A and its processing time are less than or equal to the start time of B (i.e., A is processed before B), while the disjunct on the right states that the start time of B and its processing time are less than or equal to the start time of A (i.e., B is processed before A). Clearly the OR operator, $\vee$, indicates that only one of the conditions in the disjuncts can hold true. Note that although the notation implies an inclusive OR, it is acting like an exclusive OR as only one disjunct can hold true.

7.2.1 Big-M Reformulation

Consider assigning to each disjunct a 0-1 variable y_j, such that $y_j = 1$ if the linear inequality holds true in the disjunct j, and such that $y_j = 0$ otherwise. The simplest way for reformulating the disjunction in (7.11) in terms of the binary variables y_j is with the big-M formulation given by Raman and Grossmann (1994),

$$A_j x \le b_j + M_j \left(1 - y_j \right) \quad j \in D$$

$$\sum_{j \in D} y_j = 1 \tag{7.13}$$

$$y_j = 0, 1,$$

where M_j is a sufficiently large parameter. The basic idea in the big-M constraints in (7.13) is that, for $y_j = 1$, the inequality holds true, i.e., $A_j \le b_j$. In contrast, when $y_j = 0$, the inequality becomes redundant for a sufficiently large parameter because then $A_j x \le b_j + M_j$. Clearly the value of M_j has to be carefully chosen. A large value will help to render the linear inequality to be redundant. On the other hand, a large value will cause the LP relaxation to be weak (see Section 8.1 in Chapter 8).

7.2.2 Convex-Hull Reformulation

To avoid introducing the big-M parameter, we can replace each linear disjunction in (7.11) by its convex hull (Balas, 1985; Raman and Grossmann, 1994). In this section, we present a simple derivation based on exact linearization. Later in Chapter 10, Section 10.2, we will present a formal derivation.

The disjunction in (7.11) is equivalent to the 0-1 nonlinear inequalities,

$$A_j x y_j \leq b_j y_j \quad j \in D$$
$$\sum_{j \in D} y_j = 1 \tag{7.14}$$
$$y_j = 0, 1.$$

In order to linearize the constraints in (7.14), let $z_j = xy_j$, which can be interpreted as a disaggregated variable of x. The reason is that,

$$\sum_{j \in D} z_j = \sum_{j \in D} x y_j \tag{7.15}$$

and because $\sum_{j \in D} y_j = 1$,

$$\sum_{j \in D} z_j = x, \tag{7.16}$$

which means that the variables z_j can be interpreted as disaggregated variables of the continuous variable x. To ensure that $z_j = 0$ if $y_j = 0$, we need to add the upper-bound constraints,

$$0 \leq z_j \leq U_j y_j. \tag{7.17}$$

Rewriting constraints (7.14) in terms of the variables z_j, and adding (7.16) and the upper-bound constraints in (7.17) yields the following set of mixed-integer linear constraints,

$$x = \sum_{j \in J} z_j$$
$$A_j z_j \leq b_j y_j \quad j \in D$$
$$\sum_{j \in J} y_j = 1 \tag{7.18}$$
$$0 \leq z_j \leq U_j y_j \quad j \in D$$
$$y_j = 0, 1.$$

It can be shown that the above constraints correspond to the convex hull of the disjunction in (7.11) (Balas, 1985) when the binary variables are treated as continuous, $0 \leq y_j \leq 1$.

We should note that the upper-bound constraint $0 \leq z_j \leq U_j y_j$ may not be needed if $A_i z_j \leq b_j y_j$ is such that $z_j = 0$ for $y_j = 0$. We should also note that the convex hull yields a tighter relaxation than the big-M formulation in (7.13). A simple demonstration is as follows.

Consider multiplying by y_j the big-M inequality in (7.13):

$$A_j x y_j \leq b_j y_j + M_j (1 - y_i) y_j. \tag{7.19}$$

Substituting for the variable $z_j = xy_j$, and in analogy to (7.18), we obtain,

$$A_j z_j \leq b_j y_j + M_j \left(1 - y_j\right) y_j$$

$$x = \sum_{j \in D} z_j$$

$$\sum_{j \in D} y_j = 1 \tag{7.20}$$

$$0 \leq z_j \leq U_j y_j$$

$$y_j = 0, 1.$$

Comparing (7.20) with (7.18), since $M_j \left(1 - y_j\right) y_j > 0$ for $0 < y_j < 1$, it follows that the 0-1 relaxation of the convex hull in (7.18) is tighter. On the other hand it is clear that the number of constraints and variables in the convex-hull constraints given by (7.18) is larger than those in the big-M formulation (7.13).

7.2.3 Example

Consider the following disjunction,

$$[x_1 - x_2 \leq -1] \vee [-x_1 + x_2 \leq -1]$$

$$0 \leq x_1, x_2 \leq 4. \tag{7.21}$$

Following (7.13), the big-M formulation is given by,

$$x_1 - x_2 \leq -1 + M(1 - y_1)$$

$$-x_1 + x_2 \leq -1 + M(1 - y_2)$$

$$y_1 + y_2 = 1 \tag{7.22}$$

$$y_1, y_2 = 0, 1$$

$$0 \leq x_1, x_2 \leq 4.$$

Since the continuous variables are bounded between 0 and 4 we do not need to select an excessively large value of the parameter M. For instance, the relatively modest value $M = 20$ will suffice.

Following (7.18), and for simplicity defining the variables z and w for the disaggregation of the variables x_1 and x_2, the convex-hull formulation is given by

$$x_1 = z_1 + z_2$$

$$x_2 = w_1 + w_2$$

$$z_1 - w_1 \leq -y_1$$

$$-z_2 + w_2 \leq -y_2$$

$$0 \leq z_1 \leq 4y_1$$

$$0 \leq z_2 \leq 4y_1 \tag{7.23}$$

$$0 \leq w_1 \leq 4y_2$$

$$0 \leq w_2 \leq 4y_2$$

$$y_1 + y_2 = 1$$

$$0 \leq x_1, x_2 \leq 4.$$

Figure 7.1 Relaxations of (7.22) and (7.23).

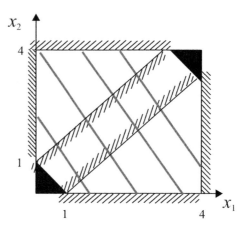

Comparing (7.22) and (7.23) it is clear that the number of constraints and variables is larger in the convex-hull formulation than in the big-M formulation.

Figure 7.1 shows the relaxed feasible regions defined by the big-M formulation (7.22) ($0 \leq x_1, x_2 \leq 4$) and the convex-hull formlation (7.23). It can be seen that the relaxed feasible region of the convex hull is tighter than that of the big-M, with the difference being the two black triangles in Fig. 7.1.

7.3   Generalized Disjunctive Programming

We close this chapter by introducing the concept of generalized disjunctive programming (GDP) (Raman and Grossmann, 1994), which is a generalization of the concept of disjunctive programming introduced by Balas (1979, 2018), which corresponds to linear programming problems with linear disjunctive constraints.

The idea in GDP is to formulate optimization problems in terms of constraints in algebraic form, and in the form of disjunctions and logic propositions. These are expressed in terms of Boolean and continuous variables. Specifically, Boolean variables Y_{jk} are assigned to each term of the disjunctions, which are represented in terms of linear constraints of continuous variables x, and continuous cost variables c_k associated with the cost of disjunction k. The GDP problem is then formulated as follows:

$$\min \quad Z = \sum_{k \in K} c_k + f^T x$$

$$\text{s.t.} \quad Dx \leq d$$

$$\bigvee_{j \in J_k} \begin{bmatrix} Y_{jk} \\ B_{jk}x \leq b_{jk} \\ c_k = \gamma_{jk} \end{bmatrix} k \in K \tag{7.24}$$

$$\underline{\bigvee_{j \in J_k}} Y_{jk}, \quad k \in K$$

$$\Omega(Y) = \text{True}$$

$$Y_{jk} = \{\text{True, False}\}, x \in X, c_k \in \mathbb{R}^1.$$

In (7.24) the inequalities $Dx \leq d$ are denoted as global constraints in that they must always be satisfied. The disjunctions $k \in K$ contain the linear inequalities $B_{jk}x \leq b_{jk}$ and the cost variables $c_k = \gamma_{jk}$ must hold true when the Boolean variable Y_{jk} = True. Note that for each disjunction k only one Boolean variable can be true due to the exclusive OR operator. The constraints $\Omega(Y)$ = True represent the logic propositions.

The GDP model in (7.24) can be used as a high-level framework for formulating discrete/continuous linear problems. Furthermore, it can also be used as a basis for deriving MILP problems (Raman and Grossmann, 1994).

One option is with big-M constraints. Here we introduce the $y_{jk} = 0, 1$ variables in place of the Boolean variables Y_{jk}. Furthermore, we replace the logic propositions $\Omega(Y)$ = True by the linear inequalities $Ay \leq a$, and substitute $c_k = \sum_{j \in I_k} \gamma_k y_{jk}$. Finally, we replace the disjunctions by the big-M constraints in (7.13). Thus, the GDP in (7.24) can be reformulated as the MILP:

$$
\begin{aligned}
\min \quad & Z = \sum_{k \in K} \sum_{j \in J_k} \gamma_k y_{jk} + f^T x \\
\text{s.t.} \quad & Dx \leq d \\
& A_j x \leq b_j + M_j \left(1 - y_{jk}\right) \quad j \in J_k, k \in K \\
& Ay \leq a \\
& \sum_{j \in D} y_{jk} = 1 \quad k \in K \\
& y_{jk} = 0, 1, \quad x \in X.
\end{aligned}
\tag{7.25}
$$

Similarly, if we replace the disjunctions by the convex-hull formulation in (7.24) this yields the following MILP, which is known as the hull reformulation (Balas, 1979, 2010, 2018; Raman and Grossmann, 1994)

$$
\begin{aligned}
\min \quad & Z = \sum_{k \in K} \sum_{j \in J_k} \gamma_k y_{jk} + f^T x \\
\text{s.t.} \quad & Dx \leq d \\
& x = \sum_{j \in J_k} z_{jk} \\
& A_j z_{jk} \leq b_j y_{jk} \quad j \in J_k, k \in K \\
& 0 \leq z_j \leq U_j y_{jk} \quad j \in J_k, k \in K \\
& Ay \leq a \\
& \sum_{j \in D} y_{jk} = 1 \quad k \in K \\
& y_{jk} = 0, 1 \quad x \in X.
\end{aligned}
\tag{7.26}
$$

We will examine generalized disjunctive programming in more detail in Chapter 10.

EXERCISES

7.1 Convert the logic expression below into a system of inequalities with 0-1 variables:

$$P_1 \vee \neg P_2 \Rightarrow P_3 \vee P_4.$$

7.2 Show that $y_1 \Rightarrow y_2$ is equivalent to $\neg y_1 \vee y_2$.

7.3 Formulate linear constraints in terms of binary variables for the following cases.

(a) If A is true and B is true then C is true or D is true (inclusive OR).

(b) The choice of all 0-1 combinations for y_j, $j \in J$ is feasible, except the one for which $y_j = 0, j \in N$, $y_j = 1, j \in B$, where N and B are specified partitions of J.

(c) If power must be generated in any period 1, 2, or 3, then install a gas turbine.

7.4 It is proposed to model the condition,

if select item 1 and not item 2, then select item 3 and item 4

with the inequality:

$$y_3 + y_4 \geq 2(y_1 - y_2),$$

where y_i are binary variables that represent the selection of the corresponding items.

Using propositional logic, derive the inequality(ies) that model the above condition. If you arrive at a different model, determine whether it is better or not, and in what sense it is better (or not) than the inequality above.

7.5 Consider a scheduling problem involving two successive time slots $k - 1$ and k and a set of N products, $i = 1, 2, \ldots, N$. We would like to model integer constraints that reflect the fact that if product i is assigned to slot $k - 1$ $(y_{i,k-1} = 1)$ and product j to slot k $(y_{j,k} = 1)$, then there is a changeover from product i to product j $(z_{ijk} = 1)$ and vice versa (see figure below).

(a) Using propositional logic, formulate linear constraints in terms of 0-1 variables to model the above condition.

(b) Show that an alternative formulation with 0-1 linear inequalities is to consider the logic fact that: (I) if there is a changeover from slot $k - 1$ with a given product i to *any product j* in slot k, then product i is assigned to slot $k - 1$, and (II) if there is a changeover from slot $k - 1$ with *any product i* to a given product j in slot k, then product j is assigned to slot k.

(c) Discuss the relative advantages of both formulations in terms of number of constraints and tightness of their relaxation.

7.6 Formulate mixed-integer linear constraints for the following disjunctions, using both big-M and convex-hull formulations.

(a) Either $0 \leq x \leq 10$ or $20 \leq x \leq 30$.

(b) The temperature approach constraint for a heat exchanger

$$T_{in} - T_{out} \geq DT \min,$$

should hold only if the exchanger is actually selected.

7.7 Convert the following generalized disjunctive programming problem into an MILP using a formulation that yields the tightest relaxation:

$$\min \quad Z = c + 2x_1 + 3x_2$$

$$\text{s.t.} \quad \begin{bmatrix} Y_1 \\ -x_1 + x_2 + 2 \le 0 \\ c = 5 \end{bmatrix} \vee \begin{bmatrix} Y_2 \\ 2 - x_2 \le 0 \\ c = 7 \end{bmatrix}$$

$$\begin{bmatrix} Y_3 \\ x_1 - x_2 \le 1 \end{bmatrix} \vee \begin{bmatrix} \neg Y_3 \\ x_1 = 0 \end{bmatrix}$$

$$Y_1 \wedge \neg Y_2 \Rightarrow \neg Y_3$$

$$\neg (Y_2 \wedge Y_3)$$

$$0 \le x_1 \le 5, 0 \le x_2 \le 5, c \ge 0$$

$$Y_j \in \{\text{true, false}\}, j = 1, 2, 3.$$

7.8 Consider the cost function shown in the graph below.
  (a) Formulate the cost function C as a disjunction.
  (b) Develop the mixed-integer constraints applying the convex hull to the disjunction.

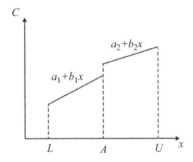

7.9 Given the nonlinear function shown below, derive (mixed-integer linear) constraints to perform a piecewise linear approximation, for fixed points x_i, $i = 0, 1, n$, and by taking linear combinations of the function values $f(x_i)$ at each interval i in terms of the weight λ, where $0 \le \lambda \le 1$.

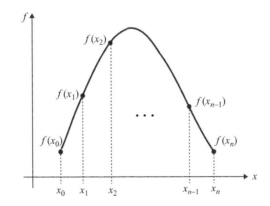

8   Mixed-Integer Linear Programming

8.1   Introduction

We consider in this chapter the solution of mixed-integer linear programming (MILP) models, which can be expressed as follows (Nemhauser and Wolsey, 1988; Wolsey, 1998):

$$\begin{aligned} \min \quad & Z = a^T x + b^T y \\ \text{s.t.} \quad & Ax + By \leq d \\ & x \geq 0, \quad y \in \{0,1\}^m, \end{aligned} \qquad \text{(MILP)}$$

in which x are the continuous variables (dimensionality n), and y are the discrete variables (dimensionality m) assumed to be 0-1 values, a very common case. In fact, any bounded discrete variable can always be reformulated in terms of 0-1 variables (see Exercise 6.1). Before we consider formal solution methods, we first outline some simple-minded solution approaches.

The simplest approach for solving problem (MILP) is a "brute force approach" in which linear programs (LP), are solved for all the 0-1 combination of the binary variables y_j, $j = 1, \ldots, m$, in order to find the global minimum. The drawback, however, is that the number of combinations increases exponentially with the number of 0-1 variables. Specifically, the number of combinations is given by 2^m. This means that for $m = 5$ there are 32 combinations, for $m = 10$ there are 10^3, while for $m = 50$ there are 10^{15} combinations. Clearly, this approach is only computationally feasible for a modest number of 0-1 variables.

Alternatively, we might consider a relaxation approach in which we relax each binary variable y_j as a continuous variable bounded between 0 and 1, that is, $0 \leq y_j \leq 1$, $j = 1, \ldots, m$. We could then consider solving the MILP as a continuous LP. Problem is that 0-1 solutions are only obtained for special structures (e.g., assignment problem) in which the matrix of coefficients is known to be totally unimodular (Nemhauser and Wolsey, 1988). Although noninteger solutions are generally obtained with this approach, the relaxed MILP yields a lower bound to the optimum solution, which will be useful in a search such as branch and bound. We should note that simply rounding the solution of the relaxed MILP to the nearest integer may yield solutions that are infeasible or suboptimal (see Fig. 8.1). Despite these limitations, it might still provide a heuristic solution that corresponds to an upper bound. Furthermore, if the rounding approach is applied to problems with integer variables that can take relatively large values, useful near-optimal solutions might be obtained.

In contrast to nonlinear programming, there are no optimality conditions that characterize the solution of the MILP. However, an elegant theory has been developed based on sequential convexification of an MILP that yields an LP whose feasible space corresponds to the integer convex hull; that is an LP where all vertices or extreme points yield 0-1 values. The sequential

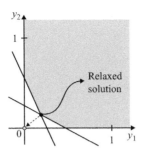

Figure 8.1 Infeasible integer rounding from LP relaxation.

convexification procedure is of course nontrivial (see Sherali and Adams, 1990; Lovasz and Schrijver, 1991; Balas et al., 1993). It is beyond of the scope of this textbook to describe in detail these convexification schemes. Suffice it to say that they roughly consist of successively multiplying the constraints by a binary variable, then linearizing the variables to lift them to a higher-dimensional space, and projecting them back to the original space. The latter step causes this approach to have exponential complexity, making it impractical except for low-dimensionality problems. However, it can be used as a framework for deriving cutting planes that can help to strengthen the LP relaxation of the MILP.

8.2   MILP Methods

The major methods for solving problem (MILP) include the following.

(a) Cutting planes. This method consists of solving a sequence of LPs in which cutting planes are successively generated to cut off the solution of the relaxed LP with previously accumulated cutting planes. The first cutting-plane method was proposed by Gomory (1958).
(b) Benders decomposition. This method proposed by Benders (1962), partitions the problem (MILP) into y and x variables, and consists of solving a sequence of integer master problems and LP subproblems. The latter generate cuts that are successively incorporated into the master problem that predicts lower bounds to the solution of the MILP.
(c) Branch and bound search. This method consists of representing the 0-1 combinations of the MILP through a binary tree in which LP subproblems are solved to determine lower and upper bounds in order to find the optimal solution by enumerating a subset of the nodes of the tree. The first branch and bound method was proposed by Land and Doig (1960), and subsequently formalized by Dakin (1965).
(d) Branch and cut methods. These methods combine the branch and bound method with cutting planes with the goal of strengthening the lower bound in the tree search (minimization case). Some of the initial branch and cut methods were described by Crowder et al. (1983) and by Van Roy and Wolsey (1987) who relied on specialized cutting planes for the knapsack and network-flow problems. The first general branch and cut method was proposed by Balas et al. (1993), who generated disjunctive cuts derived for the general (MILP) problem.

Below, we will describe in some detail the Gomory cutting planes and the branch and cut methods.

8.3 Gomory Cutting Planes

We restrict the presentation to the pure integer program given by

$$
\begin{aligned}
\min \quad & Z = c^T y \\
\text{s.t.} \quad & Ay = b \\
& y \in Z_+^n
\end{aligned}
\tag{8.1}
$$

in which the y variables correspond to nonnegative integer variables.

Assume that the LP relaxation of (8.1) is solved with $y \geq 0$ using the Simplex algorithm. Let the optimal basic variables be $y_{B_i} \in B$, and the nonbasic variables be $y_j \, j \in NB$. Problem (8.1) can then be rewritten as

$$
\begin{aligned}
\min \quad & Z = a_{00} + \sum_{j \in NB} \bar{a}_{0j} y_j \\
\text{s.t.} \quad & y_{Bi} + \sum_{j \in NB} \bar{a}_{ij} y_j = \bar{a}_{io} \quad i = 1, \ldots, m \\
& y_j \geq 0 \quad j \in NB.
\end{aligned}
\tag{8.2}
$$

From (8.2), if we round-down the right-hand side and the coefficients $\bar{a}_{ij}$ for a given noninteger row i, this yields the following inequality

$$
y_{Bi} + \sum_{j \in NB} \lfloor \bar{a}_{ij} \rfloor y_j \leq \lfloor \bar{a}_{io} \rfloor,
\tag{8.3}
$$

which is known as the Gomory cut. The question that arises is: why is (8.3) a cut? Let us substitute the equality constraint in (8.2) into (8.3). This then yields the following inequality

$$
\bar{a}_{io} - \lfloor \bar{a}_{io} \rfloor \leq \sum_{j \in NB} \left(\bar{a} - \lfloor \bar{a}_{ij} \rfloor \right) y_j.
\tag{8.4}
$$

Since $y_j = 0$, $j \in NB$ at the relaxed optimum, (8.4) then implies $\bar{a}_{io} - \lfloor \bar{a}_{io} \rfloor \leq 0$. But $\bar{a}_{io} - \lfloor \bar{a}_{io} \rfloor > 0$. This means that the inequality (8.3) is violated at $y_j = 0$, and therefore (8.3) is a cutting plane as it makes the optimal solution of (8.2) to be infeasible. While one can generate a sequence of cutting planes, by successively solving the relaxed LP and adding the previous cutting planes, the convergence is slow. This is mainly due to the fact that the cutting planes become nearly parallel (see Fig. 8.2), which in turn also induces near-singularity in the matrix of the LP. Therefore, Gomory's cutting plane is generally not a practical method.

Figure 8.2 Generation of Gomory cutting planes.

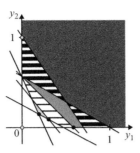

Figure 8.3 Binary tree for branch and bound search.

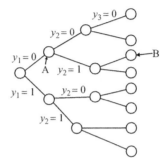

8.4 Branch and Cut Method

As indicated above, the branch and cut method is a combination of the branch and bound method with cutting planes. Major commercial codes such as CPLEX, GUROBI, and XPRESS, implement these methods, as do some of the noncommercial codes (e.g., SCIP).

The major objective in the branch and cut method is to avoid the exhaustive enumeration of a binary tree in which LP subproblems are solved at each node.

The branch and cut method consists of the two following major steps.

(a) The LP relaxation is solved at the root node of the binary tree which yields a lower bound to the MILP in (8.1). In order to strengthen this lower bound, a sequence of relaxed LPs are solved by adding cutting planes such as Gomory's. The sequence of these problems is solved until there is no improvement in the lower bound within a given tolerance ε.

(b) The branch and bound enumeration of a binary tree (see Fig. 8.3) is performed in which the problem is partitioned into a sequence of LPs, in which lower and upper bounds are generated to eliminate combinations.

We should note that each node requires solution of a relaxed LP with constraints on the binary variables. For example, in Fig. 8.3 the root node corresponds to the LP relaxation of

Figure 8.4 Hypercube for three binary variables (see text).

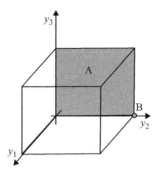

problem (MILP). Node A in the binary tree corresponds to the relaxed LP with the constraint $y_1 = 0$. Node B corresponds to the relaxed LP with the constraints $y_1 = 0$, $y_2 = 1$, $y_3 = 0$.

We should also note that the total number of nodes in the binary tree is $2^{m+1} - 1$, where m is the number of binary variables. This means that with the three binary variables the tree has 15 nodes although there are only eight binary combinations (i.e., 2^m). The hope (or gamble) is that when we enumerate the nodes in the tree, we will examine only a subset of the total number of nodes. We should also note that the sequence of LPs can be updated very efficiently with the dual Simplex algorithm with which one may require less computational time even if we have to enumerate 2^m nodes.

To appreciate why the binary tree can be interpreted as a partition consider node A in Fig. 8.3 in which the relaxed LP is solved with the constraint $y_1 = 0$. As can be seen in Fig. 8.4, that corresponds to a face of the unit cube. If we next consider node B, in which the relaxed LP is fixed with $y_1 = 0$, $y_2 = 1$, $y_3 = 0$, that corresponds to the point in the hypercube which is reached through the line $y_1 = 0$, $y_2 = 1$.

In order to derive criteria for rejecting or fathoming nodes in the enumeration of the binary tree, consider the two corresponding LP subproblems in nodes k and i, in the binary tree, where node k is a descendant of node i (e.g., $k = B$, $i = A$).

$$
\begin{aligned}
P^i : \quad & \min \ Z = a^T x + b^T y \qquad & P^k : \quad & \min \ Z = a^T x + b^T y \\
\text{s.t.} \quad & Ax + By \le d & \text{s.t.} \quad & Ax + By \le d \\
& x \ge 0, 0 \le y_j \le 1 & & x \ge 0, 0 \le y_j \le 1 \\
& y_j = 0, 1, \quad j \in J(i) & & y_j = 0, 1, \quad j \in J(k),
\end{aligned}
\tag{8.5}
$$

where $J(i)$, $J(k)$ are index sets of those binary variables with fixed 0 or 1 values. Since node i is a parent of node k, it follows that $J(i) \subset J(k)$; that is, $J(i)$ is contained in the set $J(k)$. This in turn means that problem P^k is more constrained than problem P^i. Hence, the following properties hold.

(1) If P^i is infeasible, then P^k is infeasible. This trivially follows from the fact that if P^i is infeasible, adding more constraints leading to P^k will also make this problem infeasible.

(2) If P^k is feasible then P^i is feasible and $Z^{i^*} \le Z^{k^*}$. This means that Z^{i^*}, the optimal solution of P^i, corresponds to a lower bound of the solution Z^{k^*} of P^k. This readily follows from the fact that the optimal solution of P^k must be at least as large as P^i since it is a more constrained problem.

(3) If a feasible integer solution is found for problem P^k (i.e., $y_j = 0$ or $1, j = 1, \ldots, m$) then $Z^{k^*} \ge Z^*$. This means that a feasible MILP solution at any node k, Z^k corresponds to an upper bound UB of the global optimum Z^*.

The three properties above can be used for fathoming nodes in the enumeration of the binary tree search. In particular, if subproblem P^i is infeasible then its corresponding node i can be fathomed. Similarly, if $Z^{i^*} > $ UB, where UB is the current upper bound, then node i can be fathomed.

In order to complete the branch and bound search, we need to decide how to enumerate the nodes in the tree. Since there are various ways of doing this, we consider below the common branching rules used in the branch and bound search.

The first rule addresses the question of which binary variable should be fixed to 0-1 at a given node. Note that in Fig. 8.3 we decided to enumerate the nodes in the order y_1, y_2, and y_3. But, in general, we might consider the following three schemes.

(1) Priority rule, where we specify an order for the binary variables and branch on the first noninteger binary variable; that is, we select the first $y_j \ne 0, 1$. This order usually reflects the hierarchical importance of decisions (e.g., investment vs. expansion of a plant). Consider the case where we specify the order 1, 2, 3, and that the relaxation at the root node yields $y_1 = 1$, $y_2 = 0.8$, $y_3 = 0.5$. We then select y_2 as the variables on which we generate the first branch.

(2) Rule: Select a noninteger binary closest to 0.5. The motivation here is first to resolve those variables that are the furthest from integrality. In our example above, that would mean branching first on y_3 as it has a value of 0.5, which is further away from integrality than $y_2 = 0.8$.

(3) Rule: Select a binary variable with the largest degradation (Beale, 1979). The basic idea in this approach is to estimate the down and up penalties, D_j, U_j, when forcing each noninteger binary variable y_j, to be set to 0 or to 1 as shown in Fig. 8.5. These penalties can be readily calculated with dual Simplex iterations. For each binary variable y_j the

Figure 8.5 Estimation of down and up penalties, D_j, U_j.

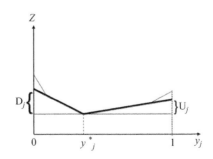

degradation T_j is set to its smallest value, $T_j = \min\{D_j, U_j\}$. The variable j to be branched on is the one with the largest degradation, that is $\max_j \{T_j\}$.

The second rule addresses the question of which node to examine next. The two major options for the tree enumeration (Morrison et al., 2016) are the following:

(a) Depth first (LIFO, newest bound rule), where the most recently created node is expanded, and backtrack takes place at terminal nodes.
(b) Best bound rule (breadth first, priority rule), where the node with best lower bound is expanded next.

For MILP, a combination of both depth first and breadth first is commonly used. Specifically, depth first is used but two nodes are expanded at each step by forcing y_j to 0 and to 1; the node with the smaller bound is selected next.

To illustrate with a small example the logic of the branch and bound search, we consider the following IP problem:

$$\min \quad Z = y_1 + 2y_2 + 4y_3$$
$$\text{s.t.} \quad y_1 + y_2 + y_3 \geq 1$$
$$y_1 - y_2 - y_3 \leq 0 \tag{8.6}$$
$$y_1 \geq y_2 + 0.2$$
$$y_1, y_2, y_3 = 0, 1.$$

As can be seen in Fig. 8.6, at node 1 the relaxed LP is solved in which the binary variables are treated as continuous. The relaxed LP yields a lower bound of $Z = 1.9$, with noninteger values for the binaries $y_1 = 0.5$, $y_2 = 0.3$, $y_3 = 0.2$. Since y_1 is the value closest to 0.5, we branch on y_1 setting it to 0 and to 1, giving rise to nodes 2 and 3, respectively. Although node 2 gives rise to an integer value $y_1 = 0$, $y_2 = 0$, $y_3 = 1$, it is infeasible. Hence, node 2 is fathomed. Node 3 on the other hand, yields a solution that is noninteger $y_1 = 1$, $y_2 = 0.8$, $y_3 = 0.2$ with an objective of $Z = 3.4$, which represents a lower bound on the solutions of the children nodes. In this case, we could either branch

Figure 8.6 Branch and bound search for example problem.

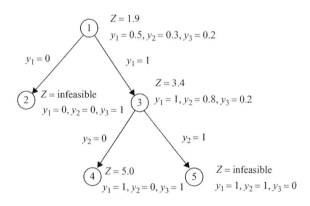

on y_2 or y_3. Say we branch on y_2. Setting it to $y_2 = 0$ yields a feasible integer solution $y_1 = 1$, $y_2 = 0$, $y_3 = 1$ at node 4 with an objective value of $Z = 5.0$. Since this solution is integer feasible it becomes an upper bound to the optimum of problem (8.6). Setting $y_2 = 1$ at node 5 yields an infeasible integer solution. Hence, node 5 is fathomed, proving that the optimum integer solution is at node 4, namely $Z = 5.0$, $y_1 = 1$, $y_2 = 0$, $y_3 = 1$.

We should note that commercial MILP codes such as CPLEX, GUROBI, and XPRESS, have become highly sophisticated in the implementation of branch and bound methods (Bixby and Rothberg, 2007). First, they incorporate a variety of cutting planes, such as Gomory's cutting planes, in order to strengthen the lower bounds predicted at the nodes of the branch and bound search. Typically, most of these cutting planes are added at the root node until the improvement of the lower bound lies within a specified tolerance. The main effect of the cutting planes is then to reduce the number of nodes in the branch and bound search. Another major impact in MILP codes has been the use of preprocessing techniques that allow a size reduction of models through the use of algebraic substitutions and logic inference and constraint propagation based on constraint programming techniques (see Chapter 11).

See Appendix A for further discussion on MILP codes.

EXERCISES

8.1 Given is the integer programming problem

$$\max \quad Z = 1.2y_1 + y_2$$
$$\text{s.t.} \quad y_1 + y_2 \leq 1$$
$$1.2y_1 + 0.5y_2 \leq 1$$
$$y_1, y_2 = 0, 1.$$

(a) Plot the contours of the objective and the feasible region for the case when the binary variables are relaxed as continuous variables $y_1, y_2 \in [0, 1]$.

(b) Determine from inspection the solution of the relaxed problem.

(c) Enumerate the four 0-1 combinations in your plot to find the optimal solution.

(d) Solve the relaxed LP problem by hand and derive Gomory cuts based on the LP relaxation and verify that they cut off the relaxed LP solution.

(e) Solve the above problem with the branch and bound method by enumerating the nodes in the tree and solving the LP subproblems with GAMS.

8.2 (a) Show that the number of nodes in a tree where we represent all possible combinations of m 0-1 binary variables is $2^{m+1} - 1$.

(b) If a complete enumeration of all the nodes in the tree were required, by what factor would this enumeration increase with respect to the direct enumeration of all 0-1 combinations?

8.3 It is desired to model a condition that states that if any of n items i are to be transported (y_i), then one must buy a container for them $(z = 1)$. Determine if an alternative constraint can be formulated that is tighter than

$$\sum_i y_i \le nz, \quad y_i = 0, 1, \quad i = 1, 2, \ldots, n, \quad z = 0, 1.$$

8.4 When solving an MILP problem you find that the relaxed LP has a finite optimum at the root node, but that the descendants' nodes produce infeasible LP subproblems. Explain why this may be happening.

8.5 A company is considering the production of a chemical C, which can be manufactured with either process II or process III, both of which use chemical B as a raw material. Chemical B can be purchased from another company, or else manufactured with process I, which uses A as a raw material. Given the specifications below, formulate an MILP model and solve it with GAMS to reach a decision.

(a) Which process should be built (II and III are exclusive)?

(b) How should we obtain chemical B?

(c) How much of product C should be produced?

The objective is to maximize profit.

Consider the two following cases.

(1) The maximum demand of C is 10 tons/hr, with a selling price of $1800/ton.

(2) The maximum demand of C is 15 tons/hr; the selling price for the first 10 ton/hr is $1800/ton, and $1500/ton for the excess.

Data

	Investment and operating costs	
	Fixed ($/hr)	Variable ($/ton raw mat.)
Process I	1000	250
Process II	1500	400
Process III	2000	550

Prices	A	$500/ton
	B	$950/ton

Conversions	Process I	90% of A to B
	Process II	82% of B to C
	Process III	95% of B to C

The maximum supply of A is 16 tons/hr.

NOTE: You may want to scale your cost coefficients (e.g., divide them by 100).

9   Mixed-Integer Nonlinear Programming

9.1   Overview of Solution Methods

We consider next the case where the mixed-integer programming problem involves non-linearities. In general, the MINLP problem with 0-1 discrete variables can be formulated as follows (Floudas, 1995; Grossmann, 2002; Trespalacios and Grossmann, 2014):

$$\begin{aligned} \min \quad & Z = f(x, y) \\ \text{s.t.} \quad & h(x, y) = 0 \\ & g(x, y) \leq 0 \\ & x \in R^n, \quad y \in \{0, 1\}^m. \end{aligned} \quad \text{(MINLP)}$$

While problem (MINLP) is quite general, it is convenient to consider the special structure in which the 0-1 variables appear in linear form in the objective function and constraints, while the continuous variables are in general nonlinear. Furthermore, we specify the set X of continuous variables as having linear constraints with lower and upper bounds, x^L, x^U; that is $X = \{x | Cx \leq d, x^L \leq x \leq x^U\}$. In this way, problem (MINLP) is given by

$$\begin{aligned} \min \quad & Z = c^T y + f(x) \\ \text{s.t.} \quad & h(x, y) = 0 \\ & g(x) + By \leq 0 \\ & Ay \leq a \\ & x \in X, \quad y \in \{0, 1\}^m, \end{aligned} \quad \text{(MINLP}')$$

in which the linear structure can be exploited as will be shown below. The two major approaches for solving problem (MINLP') are the following.

(1) <u>Branch and bound</u>. This method (Gupta and Ravindran, 1985) is similar to solving problem mixed-integer linear programs (MILP) in the previous chapter. The major difference, however, is that at each node, relaxed NLP subproblems are solved with corresponding constraints for the fixed subset of 0-1 variables corresponding to that node. This solution approach is reasonable when the number of 0-1 variables is not too large, as otherwise one might face the solution of many NLP subproblems, which are computationally more expensive than solving LP subproblems in the MILP case. If the NLP relaxation of the MINLP is tight, larger problems can be solved with the NLP-based branch and bound. Codes that implement this method for MINLP include SBB and BONMIN.

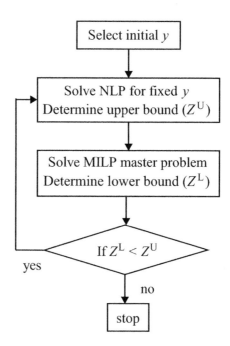

Figure 9.1 Steps of decomposition algorithms for MINLP.

(2) Decomposition. These methods consist of iterating between NLP subproblems with fixed binary variables, and MILP master problems that predict new values for the binary variables. The NLP subproblems provide upper bounds Z^U if there are feasible solutions, while the MILP subproblems predict lower bounds Z^L. Iterations are continued until the bounds lie within a given tolerance, as shown in Fig. 9.1.

Methods that use this solution approach include generalized Benders decomposition (Geoffrion, 1972), outer approximation (Duran and Grossmann, 1986), LP/NLP-based branch and bound (Quesada and Grossmann, 1992), and extended cutting plane (Westerlund and Petersson, 1995). Codes that implement these methods for solving MINLP problems are DICOPT, α-ECP, and BONMIN.

9.2 Derivation of Outer-Approximation and Generalized Benders Decomposition Methods

For the derivation of decomposition methods we assume that there are no equations $h(x) = 0$ in the MINLP′ case, and that the functions $f(x), g(x)$ are convex. This leads to the MINLP in (9.1), which is denoted as convex MINLP since its NLP relaxation is convex:

$$\begin{aligned}
\min \quad & Z = c^T y + f(x) \\
\text{s.t.} \quad & g(x) + By \le 0 \\
& Ay \le a \\
& x \in X, \quad y \in \{0,1\}^m.
\end{aligned} \tag{9.1}$$

The subproblem at a fixed value y^k of the binary variables is given by:

$$Z(y^k) = \min c^T y^k + f(x)$$
$$\text{s.t.} \quad g(x) + By^k \leq 0 \tag{$S(yk)$}$$
$$x \in X.$$

If the subproblem we have denoted $(S(y^k))$ has a feasible solution, its optimal objective yields an upper bound to MINLP in (9.1). If it has no feasible solution it is convenient to define the feasibility problem:

$$\min \quad \eta$$
$$\text{s.t.} \quad g(x) + By^k \leq e\eta \tag{$F(yk)$}$$
$$x \in X, \ \eta \in R^1,$$

where e is the unit vector.

For the master problem, develop an equivalent linear representation of the MINLP in (9.1) and solve it by relaxation. Specifically, we assume $f(x)$, $g(x)$ are convex differentiable functions. Next, by transferring the objective as a constraint with the variable α, we rearrange the MINLP in (9.1) as follows:

$$Z = \min \alpha$$
$$\text{s.t.} \quad \alpha \geq c^T y + f(x)$$
$$g(x) + By \leq 0 \tag{9.2}$$
$$Ay \leq a$$
$$x \in X, \alpha \in R^1 \quad y \in \{0,1\}^m.$$

Because $f(x)$, $g(x)$ are convex, first-order linearizations at points x^k yield supporting hyperplanes at each of these points, namely,

$$\left. \begin{array}{l} f(x) \geq f(x^k) + \nabla f(x^k)^T (x - x^k) \\ g(x) \geq g(x^k) + \nabla g(x^k)^T (x - x^k) \end{array} \right\} \forall x^k \in X. \tag{9.3}$$

Furthermore, the linearizations at the infinite set of points provides a valid representation for the functions $f(x)$ and $g(x)$. For a finite set of points the linearizations provide an underestimation to the objective function and an overestimation to the feasible region as shown in Figs. 9.2a and 9.2b.

If we substitute the linearizations (9.3) into (9.2) this yields the MILP master problem (MOA):

$$Z = \min \alpha$$
$$\text{s.t.} \quad \left. \begin{array}{l} \alpha \geq c^T y + f(x^k) + \nabla f(x^k)^T (x - x^k) \\ g(x^k) + \nabla g(x^k)^T (x - x^k) + By \leq 0 \\ Ay \leq a \end{array} \right\} \forall x^k \in T \tag{9.4}$$
$$x \in X, \alpha \in R^1 \quad y \in \{0,1\}^m$$
$$T = \{x^k | \text{is solution of } S(y^k) \text{ or } F(y^k), \forall y^k\}.$$

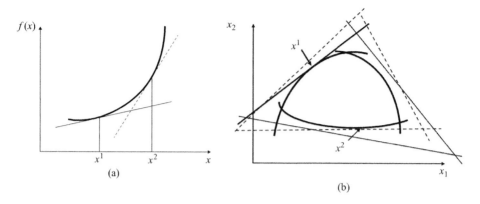

Figure 9.2 (a) Linearizations of the convex function underestimate the objective function, and (b) overestimate the feasible region.

This master problem has the following property as stated in Theorem 9.1.

Theorem 9.1 *Problems MINLP and MOA have the same optimal solution* (x^*, y^*).

The proof trivially follows from the fact that the functions $f(x)$ and $g(x)$ can be replaced by the infinite set of linearizations in (9.3). Because it is not practical to obtain linearizations at an infinite number of points, we assume we are given the solution of K NLP subproblems x^k at y^k, $k = 1, \ldots, K$. We can then define the relaxed MILP problem RM_{OA}^K for the outer-approximation (OA) method:

$$
Z_{\text{OA}}^K = \min \alpha_{\text{OA}}
$$

$$
\left.
\begin{array}{l}
\alpha_{\text{OA}} \geq c^T y + f\left(x^k\right) + \nabla f\left(x^k\right)^T \left(x - x^k\right) \\[1mm]
\text{s.t.} \quad g\left(x^k\right) + \nabla g\left(x^k\right)^T \left(x - x^k\right) + By \leq 0 \\[1mm]
Ay \leq a
\end{array}
\right\} \quad k = 1, \ldots, K
\tag{9.5}
$$

$$
x \in X, \quad \alpha_{\text{OA}} \in R^1, y \in \{0, 1\}^m.
$$

Generalized Benders decomposition (GBD) involves an MILP master problem that corresponds to a projection in the space of the binary variables y. This is accomplished through a dual representation in terms of the binary variables y that can be derived as follows.

Consider an outer approximation at y^k in MOA

$$
\alpha \geq c^T y + f\left(x^k\right) + \nabla f\left(x^k\right)^T \left(x - x^k\right),
\tag{9.6}
$$

$$
g\left(x^k\right) + \nabla g\left(x^k\right)^T \left(x - x^k\right) + By \leq 0.
\tag{9.7}
$$

Let $\mu^k \geq 0$ be the optimal multiplier of the inequality of $g(x) + By \leq 0$ from subproblem $S(y^k)$ at the point y^k. From (9.7), by premultiplying by μ^k we obtain

$$\left(\mu^k\right)^T\left[g\left(x^k\right) + By\right] \leq -\left(\mu^k\right)^T \nabla g\left(x^k\right)^T\left(x - x^k\right). \tag{9.8}$$

From the KKT conditions of subproblem $(S(y^k))$ and, for simplicity, assuming we have no active variable bounds, we have,

$$\nabla f\left(x^k\right) + \nabla g\left(x^k\right)\mu^k = 0. \tag{9.9}$$

If we postmultiply (9.9) by $(x - x^k)$, and rearranging the equality, we have,

$$\nabla f\left(x^k\right)^T\left(x - x^k\right) = -\left(\mu^k\right)^T \nabla g\left(x^k\right)^T\left(x - x^k\right). \tag{9.10}$$

Substituting (9.10) into (9.8) yields

$$\left(\mu^k\right)^T\left[g\left(x^k\right) + By\right] \leq \nabla f\left(x^k\right)^T\left(x - x^k\right). \tag{9.11}$$

Finally, if we substitute into (9.6), we obtain the Benders cut,

$$\alpha \geq c^T y + f\left(x^k\right) + \left(\mu^k\right)^T\left[g\left(x^k\right) + By\right]. \tag{9.12}$$

This cut, which is a surrogate of the constraints (9.6) and (9.7), corresponds to a projection of the outer approximations (9.6) and (9.7) into the space (α, y). The cut can also be interpreted as the Lagrangean function parametrized in the variables y,

$$\mathcal{L}(y) = c^T y + f\left(x^k\right) + \left(\mu^k\right)\left[g\left(x^k\right) + By\right]. \tag{9.13}$$

For the case that the subproblem $(S(y^k))$ has no feasible solution, the feasibility cut that can be derived from the feasibility problem $(F(y^k))$ is given by

$$\left(\mu^k\right)^T\left[g\left(x^k\right) + By\right] \leq 0. \tag{9.14}$$

If we replace (9.6) and (9.7) by (9.12) and (9.14), and given the solution of K NLP subproblems at y^k, $k = 1, \ldots, K$ with solution x^k, μ^k, we can define the relaxed MILP master problem RM_{GB}^K for GBD,

$$
\begin{aligned}
Z_{\text{GB}}^K = \ & \min \alpha_{\text{GB}} \\
\text{s.t.} \quad & \alpha_{\text{GB}} \geq c^T y + f\left(x^k\right) + \left(\mu^k\right)^T\left[g\left(x^k\right) + By\right] \quad k \in K_F^K \\
& \left(\mu^k\right)^T\left[g\left(x^k\right) + By\right] \leq 0 \qquad\qquad\quad k \in K_I^K \\
& \alpha_{\text{GB}} \in R^1 \quad y \in \{0, 1\}^m,
\end{aligned}
\tag{9.15}
$$

where $k \in K_F^K$ and $k \in K_I^K$ correspond to the optimality and feasibility cuts, respectively.

Based on the above, we can make the following remarks.

(1) RM_{OA}^K and RM_{GB}^K are MILP master problems that accumulate linearizations as iterations $k = 1, \ldots, K$ proceed.
(2) The optimal solutions to RM_{OA}^K and RM_{GB}^K correspond to lower bounds $Z_{\text{OA}}^K, Z_{\text{GB}}^K$ in (9.5) and (9.15), respectively. These lower bounds increase monotonically with the iterations k, that is (see Fig. 9.3),

Figure 9.3 Progression of lower and upper bounds in OA and GBD methods.

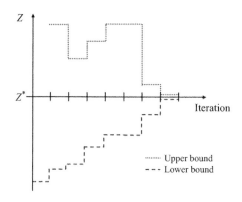

$$Z^1_{\text{OA}} \leq Z^2_{\text{OA}} \leq \cdots \leq Z^K_{\text{OA}}$$
$$Z^1_{\text{GB}} \leq Z^2_{\text{GB}} \leq \cdots \leq Z^K_{\text{GB}}.$$

(3) The OA method predicts stronger lower bounds than GBD, $Z^K_{\text{OA}} \geq Z^K_{\text{GB}}$.

This follows from the fact that the Lagrangean cut in (9.12) is a linear combination or surrogate of the linearization in (9.6) and (9.7) of the OA method. Furthermore, this implies that the OA method generally requires fewer iterations than the GBD method. The trade-off is that the OA involves a larger MILP master problem, (9.5) vs. (9.15), which is more expensive to solve.

9.3 Extended Cutting-Plane Method

The extended cutting-plane (ECP) method of Westerlund and Petersson (1995) is similar to the OA algorithm, except that it replaces the NLP step in OA by a simple function evaluation, which is used to provide a new linearization to an MILP master problem of the form of (9.5). In this way the ECP method resembles a successive MILP procedure. The iterations continue until the nonlinear functions and constraints are converged within a tolerance ε. The specific steps of the ECP are then as follows.

(1) Guess x^0, y^0 set $K = 0$.
(2) Find x^{K+1}, y^{K+1} from the master problem of ECP, RM^K_{ECP}.

$$\min \ Z = \alpha_{\text{ECP}}$$

$$\left.
\begin{aligned}
& \alpha_{\text{ECP}} \geq c^T y + f\left(x^k\right) + \nabla f\left(x^k\right)^T \left(x - x^k\right) \\
\text{s.t.} \quad & g\left(x^k\right) + \nabla g\left(x^k\right)^T \left(x - x^k\right) + By \leq 0 \\
& Ay \leq a
\end{aligned}
\right\} k = 1, \ldots, K.$$

$$x \in X, \quad \alpha_{\text{ECP}} \in R^1, \quad y \in \{0, 1\}^m.$$

(3) If $\left\| g\left(x^{k+1} \right) + By^{k+1} \right\| \leq \varepsilon$, STOP.

Otherwise, set $K = K + 1$, linearize functions at $\left(x^k, y^k \right)$, and return to step (2).

We should note that because the ECP method is expected to require a larger number of iterations, a larger number of successive MILP master problems must be solved. Instead of linearizing all the nonlinear constraint functions $g(x)$, usually only the functions corresponding to the most violated inequalities are linearized.

A recent extension of the ECP method is the extended supporting-hyperplane (ESH) method (Kronqvist et al., 2016). The main idea is not simply to linearize the functions at the solution of the master problem of the ECP (step (2) above) to obtain supporting hyperplanes at each iteration, and use these to construct a polyhedral outer approximation. Specifically, a strict interior point $\left(x^{\text{int}}, y^{\text{int}} \right)$ is obtained from the NLP subproblem,

$$
\begin{aligned}
\min \quad & \mu \\
\text{s.t.} \quad & g_j(x, y) \leq \mu, \qquad j \in J \\
& x \in X, y \in \{0, 1\}^m, \quad \mu \in R^1.
\end{aligned} \tag{9.16}
$$

To construct the polyhedral outer approximation, a new function F is defined as the pointwise maximum of the nonlinear constraints, according to

$$
F(x, y) = \max_{j \in J} \left\{ g_j(x, y) \right\}. \tag{9.17}
$$

The new points $\left(x^k, y^k \right)$ are then determined by

$$
\begin{aligned}
x^k &= \lambda^k x^{\text{int}} + \left(1 - \lambda^k \right) x^{\text{MIP}} \\
y^k &= \lambda^k y^{\text{int}} + \left(1 - \lambda^k \right) y^{\text{MIP}},
\end{aligned} \tag{9.18}
$$

where x^{MIP} and y^{MIP} are the solution of the master MILP in step (2), and λ^k is chosen such that $F(x, y) = 0$, i.e., at the boundary of the feasible region. The points $\left(x^k, y^k \right)$ are then located at the boundary of the feasible region, so that linearizing the active constraints results in supporting hyperplanes. Numerical experience has shown that these linearizations significantly improve the convergence of the ECP method.

9.4   Properties and Extensions

The following properties can be established for the OA, GBD, and ECP methods presented in the two previous sections.

(1) We should note that the OA and GBD methods have finite convergence, because either they terminate by equality of the lower and upper bounds or, in the worst case, they would end up enumerating all possible binary combinations, 2^m.

(2) The branch and bound method converges in one iteration when the NLP relaxation in (9.5), in which the binary variables are treated as continuous variables, $0 \leq y_i \leq 1$, $i = 1, \ldots, m$, yields an integer solution. This effectively means that the MINLP problem is solved at the root node of the tree.

(3) The OA and ECP methods converge in one iteration for the case when the functions $f(x)$ and $g(x)$ in the MINLP (9.1) are linear. This trivially follows from the fact that, in that case problem (9.1), an MILP, becomes identical to the MILP master problems, RM_{OA}^K and RM_{ECP}^K.

(4) Sahinidis and Grossmann (1991) proved that the GBD method converges in one iteration for optimal y^* and if the problem (MINLP) in (9.5) has zero integrality gap. This property points to the importance of developing tight MINLP models when applying the GBD method.

A recent review of solvers for convex MINLP problems can be found in Kronqvist et al. (2019).

In the previous sections we have addressed the case of when the MINLP problem in (9.5) does not involve equations $h(x) = 0$. The difficulty is that, generally, nonlinear equations give rise to nonconvexities, unless they can be shown to relax as convex inequalities. Specifically, we would require the relaxations of the equations to satisfy the following conditions,

$$h_i(x) = \begin{cases} h_i(x) \leq 0 & \text{convex} \\ h_i(x) \geq 0 & \text{concave.} \end{cases} \tag{9.19}$$

Assuming we apply the algorithms without regard as to whether convexity applies or not, the extensions for addressing problems of the form of (MINLP') in Section 9.1 with non-linear equations are as follows.

– In GBD it is easy if the NLP subproblems are feasible. The reason is that analyzing the Lagrangean function $\mathcal{L} = c^T y + f(x^k) + (\lambda^k)^T h(x^k) + (\mu^k)^T [g(x^k) + By]$, it can be seen that, if $h(x^k) = 0$, the Lagrangean reduces to the GBD cut given in (9.12).

– In the OA algorithm, because we cannot simply linearize the equations and add them to the previous ones, as this would give rise to overspecifications and invalid cuts, we can resort to relaxing the equations $h(x) = 0$ by analyzing their KKT multipliers λ_i of the corresponding NLP problem as follows (Kocis and Grossmann, 1987).

Relax as $h_i(x) \leq 0$ if $\lambda_i > 0$.
Relax as $h_i(x) \geq 0$ if $\lambda_i < 0$.
If $\lambda_i = 0$, remove $h_i(x) = 0$.

Based on the above, we can define the relaxation matrix $T^k = \{t_{ii}^k\}$

$$t_{ii}^k = \begin{cases} 1 & \text{if } \lambda_i > 0 \\ -1 & \text{if } \lambda_i < 0 \\ 0 & \text{if } \lambda_i = 0. \end{cases} \tag{9.20}$$

In this way, for feasible points for which $h(x^k) = 0$, the linearizations can be expressed as follows:

Figure 9.4 Cutting off global optimum with linearization of nonconvex function.

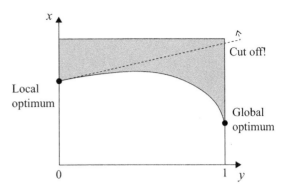

$$T^k \left[\nabla h(x^k)^T (x - x^k) \right] \leq 0. \tag{9.21}$$

We should note that one can also add an integer cut to make integer points y^k that have been examined at previous iterations $k = 1, \dots, K$ to be infeasible. This can be accomplished with the inequalities (see Exercise 6.2),

$$\sum_{i \in B^k} y_i - \sum_{i \in N^k} y_i \leq |B^k| - 1 \quad k = 1, \dots, K, \tag{9.22}$$

where $B_i^k = \{ i | y_i^k = 1 \}$, $\quad N_i^k = \{ i | y_i^k = 0 \}$.

As for the issue of nonconvexity, we should note that convexity is a sufficient condition for the OA, GBD, ECP, and branch and bound methods to converge to the global optimum. Difficulties that may arise due to nonconvexities are that the NLP subproblems may have multiple solutions. This implies that the upper bounds may not be valid. Furthermore, in the OA, GBD, ECP methods nonconvex functions may cut off the optimum in the MILP master problem since the linearizations may not underestimate the feasible region as is shown in the simple example in Fig. 9.4.

Two options to overcome these problems include trying to convexify the MINLP problem, or applying rigorous global-optimization methods, as will be discussed in Chapter 12 on global optimization. The other option is to introduce slacks with an augmented penalty in the spirit of SQP algorithms (Powell, 1978) in order to allow for the violation of nonconvex constraints.

In the case of the MILP master problem of the OA method, we can modify it by implementing equality relaxation, introducing slack variables for an augmented penalty, and adding integer cuts. The modified MILP master problem is then as follows (Viswanathan and Grossmann, 1990).

Let ρ be the penalty weight for the slack variables $s^k, p^k, q^k \geq 0$. The MILP master $\text{RM}_{\text{OA}}^{\text{AP}}$ of equality relaxation with augmented penalty is given by

$$\min \quad Z_{OA}^K = \alpha_{OA} + \rho \sum_k \left(s^k + e^T p^k + e^T q^k \right)$$

$$\text{s.t.} \quad \left. \begin{array}{l} c^T y + f\left(x^k\right) + \nabla f\left(x^k\right)^T \left(x - x^k\right) - \alpha_{OA} \leq s^k \\[2mm] T^k \left[\nabla h\left(x^k\right)^T \left(x - x^k\right) \right] \leq p^k \\[2mm] g\left(x^k\right) + \nabla g\left(x^k\right)^T \left(x - x^k\right) + By \leq q^k \\[2mm] Ay \leq a \end{array} \right\} k = 1, \ldots, K \qquad (9.23)$$

$$\sum_{i \in B_k} y_i - \sum_{i \in N_k} y_i \leq \left| B^k \right| - 1 \quad k = 1, \ldots, K$$

$$x \in X, \quad \alpha_{OA} \in R^1, \ s^k, p^k, q^k \geq 0 \quad y \in \{0,1\}^m.$$

We should note that if ρ is large ($\rho > |\gamma_i|$), then if the functions are convex (including relaxed equations) the slack variables $s^k = p^k = q^k = 0$, which implies that the MILP master problem in (9.23) reduces to the original MILP master problem (9.5) (except for the integer cuts).

If the functions involved in the MINLP are nonconvex, the optimal value of the MILP master problem in (9.23), Z_{OA}^K, will not correspond to a rigorous lower bound, among other things, because the slacks will generally be nonzero. However, we can resort to a heuristic termination criterion. Specifically, the termination rule is to stop until there is no improvement in the NLP subproblems. Obviously, this will generally not provide the global-optimum solution, although computational experience seems to indicate that in surprisingly many cases one does find near-optimum solutions.

The computational strategy based on the augmented-penalty MILP master problem in (9.23), and the termination criterion of nonimprovement, has been implemented in the software code DICOPT, where the relaxed NLP is initially solved in order to generate a point for the initial master problem (see Fig. 9.5). If that relaxed NLP yields an integer solution, obviously no further search needs to be conducted.

MINLP optimization has been a very active area of research in the recent past (Bonami et al., 2012). Appendix A describes some of the software for solving MINLP that has been developed in the recent past.

Figure 9.5 Steps in DICOPT solver for MINLP.

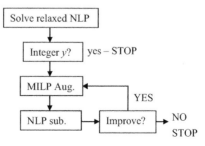

EXERCISES

9.1 Show that if $f(x)$ is a convex differentiable function, then

$$f(x) \geq f(x^k) + \nabla f(x^k)^T (x - x^k) \quad \text{for all } x.$$

9.2 For what special class of MINLP problems will the branch and bound method for solving MINLP problems outperform the OA, GBD, and ECP methods in terms of number of NLP subproblems to be solved?

9.3 Using GAMS solve the following MINLP problem step by step with
(a) generalized Benders decomposition,
(b) outer-approximation method,
(c) extended cutting plane.
(d) Also verify your answer with GAMS/DICOPT:

$$
\begin{aligned}
\min \quad & f = y_1 + 1.5y_2 + 0.5y_3 + x_1^2 + x_2^2 \\
\text{s.t.} \quad & (x_1 - 2)^2 - x_2 \leq 0 \\
& x_1 - 2y_1 \geq 0 \\
& x_1 - x_2 - 4(1 - y_2) \leq 0 \\
& x_1 - (1 - y_1) \geq 0 \\
& x_2 - y_2 \geq 0 \\
& x_1 + x_2 \geq 3y_3 \\
& y_1 + y_2 + y_3 \geq 1 \\
& 0 \leq x_1 \leq 4, \quad 0 \leq x_2 \leq 4 \\
& y_1, y_2, y_3 = 0, 1.
\end{aligned}
$$

(e) Use the starting point $y_1 = y_2 = y_3 = 1$,
$x_1 = x_2 = 0$ for extended cutting plane.

9.4 Consider the "mostly linear" mixed-integer nonlinear programming problem of the following form, where $r(v), t(v)$ are convex nonlinear functions in terms of "complicating" nonlinear continuous variables v, while w are the linear variables:

$$
\begin{aligned}
\min \quad & Z = c^T y + a^T w + r(v) \\
\text{s.t.} \quad & Cy + Dw + t(v) \leq 0 \\
& Ey + Fw + Gv \leq d \\
& y \in \{0, 1\}, w \in W, v \in V.
\end{aligned}
$$

Show that the outer approximation of the above problem at fixed value of the binary variable y^k, and corresponding optimal solution x^k, and optimal multipliers λ_k, μ_k^T, can be expressed by the following constraints in which only the nonlinear variables v are projected out by Benders cuts:

$$\alpha \geq c^T y + a^T w + r(v^k) + (\lambda_k^T)[Cy + Dw + t(v^k)] - (\mu_k^T)G(v - v^k)$$
$$Ey + Fw + Gv \leq d$$
$$y \in \{0, 1\}, w \in W, v \in V.$$

9.5 It is proposed to manufacture a chemical C with a process I that uses raw material B. B can be purchased or manufactured with either of two processes, II or III, which use chemical A as a raw material. In order to decide the optimal selection of processes and levels of production that maximize profit, formulate the MINLP problem and solve with the augmented-penalty/outer-approximation/equality-relaxation algorithm in DICOPT.

Data:

Conversion: Process I $C = 0.9B$ Process II $B = \ln(1 + A)$ Maximum capacity: 5 ton prod/hr

$\qquad\qquad$ Process III $B = 1.2 \ln(1 + A)$ $(A, B, C,$ in ton/hr$)$

Prices: $A\$ 1800/$ton
$\qquad\quad$ $B\$ 7000/$ton
$\qquad\quad$ $C\$ 13\,000/$ton (maximum demand: 1 ton/hr)

Investment cost

	Fixed $(10^3 \ \$/hr)$	Variable $(10^3 \ \$/$ton product$)$
Process I	3.5	2
Process II	1	1
Process III	1.5	1.2

Note: Minimize negative of profit.

10 Generalized Disjunctive Programming

10.1 Logic-Based Formulation for Discrete/Continuous Optimization

We now consider the case of generalized disjunctive programming (GDP) models, which are discrete and continuous optimization models that involve equations, disjunctions, and logic propositions (Raman and Grossmann, 1994; Grossmann and Trespalacios, 2013). We should note that disjunctive programming refers to linear programming with disjunctions as per the pioneering work by Balas (1979, 2018). Here we denote the model as GDP because it involves nonlinear functions in the continuous variables and, in addition to disjunctions, it contains logic propositions in terms of Boolean variables. These variables are specified in each term of the disjunction and serve as indicators as, if they are true, a subset of equations must also hold true. In Chapter 7 we introduced GDP models for linear constraints, and showed how they can be reformulated as MILP models, either through the big-M reformulation or through the hull reformulation. In this chapter we assume that the functions are generally nonlinear.

More specifically, the GDP problem is given as follows:

$$\min \quad Z = \sum_{k \in K} c_k + f(x)$$

$$\text{s.t.} \quad r(x) \leq 0$$

$$\bigvee_{j \in J_k} \begin{bmatrix} Y_{jk} \\ g_{jk}(x) \leq 0 \\ c_k = \gamma_{jk} \end{bmatrix} k \in K \qquad \text{(GDP)}$$

$$\underset{j \in \overline{J_k}}{\vee} Y_{jk}$$

$$\Omega(Y) = \text{True}$$

$$0 \leq x \leq U, \quad c_k \in R^1, \quad Y_{jk} = \{\text{True}, \text{False}\},$$

where $x \in R^n$ are the continuous variables, Y_{jk} are the Boolean variables that take values of True or False, $r(x) \leq 0$ are global inequalities that must always hold true, and K is the set of disjunctions in terms of the OR operator $\vee$, in which each term $j \in J_k$ or disjunct contains the Boolean variable Y_{jk}, the inequalities $g_{jk}(x) \leq 0$, and the cost variable c_k, which is equated to a fixed cost γ_{jk}. If the Boolean variable is True both the inequalities and the cost equation are enforced; if the Boolean variable is False they are both ignored. Thus, each disjunct is equivalent to an IF THEN ELSE condition. Because only one term of a disjunction k can be true, we enforce the exclusive OR on the Boolean variables. Next, we specify the logic propositions through the symbolic logic relations $\Omega(Y) = \text{True}$. Finally, bounds are specified on the continuous variables x and the Y_{jk} are declared as Boolean variables.

We should note that a major motivation for formulating discrete/continuous problems in the form of GDP problems is that they can help to facilitate the formulation of a model because the disjunctions and logic provide a higher level of representation that does not require the user to pose the problem in terms of only algebraic equations as is the case in mixed-integer programming. See, for instance, Castro and Grossmann (2012) for the contrast of formulations between mixed-integer and GDP models in the area of batch-scheduling optimization.

In order to address the solution of problem (GDP), we first assume, as we did for problem (MINLP), that $f(x), r(x), g_{jk}(x)$ are convex differentiable functions. We consider next three major solution approaches to solving problem (GDP) (Grossmann, 2002; Trespalacios and Grossmann, 2014).

(a) Reformulation to an MINLP problem, which can be accomplished either by formulating each disjunction through equations describing the convex hull, or through big-M constraints.
(b) Disjunctive branch and bound that uses at each node a relaxation based on taking the convex hull of each disjunction or using big-M constraints.
(c) Decomposition approach that partitions the problem into continuous NLP optimization problems and discrete MILP master problems. Two possible approaches include the logic-based outer approximation and logic-based generalized Benders decomposition.

10.2 Relaxations and Reformulations of GDP

In order to obtain an NLP relaxation of problem (GDP), we consider first the convex hull of a given disjunction k in (GDP) (Lee and Grossmann, 2000). This can be defined by considering variable points $u_{jk}, j \in J_k$ for each disjunction k such that

$$g_{jk}\left(u_{jk}\right) \leq 0 \quad j \in J_k. \tag{10.1}$$

Taking a linear combination of these points for all $\lambda_{jk} \geq 0$ leads to

$$x = \sum_{j \in J_k} \lambda_{jk} u_{jk} \quad \sum_{j \in J_k} \lambda_{jk} = 1. \tag{10.2}$$

Performing a similar linear combination for the cost variable leads to

$$c_k = \sum_{j \in J_k} \lambda_{jk} \gamma_{jk}. \tag{10.3}$$

A geometric interpretation of these linear combinations is shown in Fig. 10.1.

Note that the term $\lambda_{jk} u_{jk}$ is bilinear. In order to linearize it, let

$$v_{jk} = \lambda_{jk} u_{jk}, \tag{10.4}$$

which, from (10.2), then leads to the linear equality

$$x = \sum_{j \in J_k} v_{jk}, \tag{10.5}$$

Figure 10.1 Convex hull of $x \in S_1 \vee x \in S_2$.

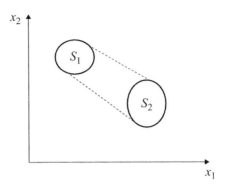

where $0 \le u_{jk} \le U\lambda_{jk}$. Because $g_{jk}(u_{jk}) \le 0$ and $u_{jk} = \frac{v^{jk}}{\lambda_{jk}}$, $\lambda_{jk} \ge 0$, the inequality constraint in (10.1) can be set to

$$\lambda_{jk} g_{jk}\left(\frac{v^{jk}}{\lambda_{jk}}\right) \le 0, \tag{10.6}$$

where we also multiply the inequality by λ_{jk}. This, at first sight, is counterintuitive, but it has the justification that it gives rise to the perspective function $\lambda g\left(\frac{v}{\lambda}\right)$, which has the following desirable property (Hiriart-Urruty and Lemaréchal, 2001):

Property 10.1 The perspective function $\lambda g(v/\lambda)$ is convex if $g(\cdot)$ convex and $\lambda g\left(\frac{v}{\lambda}\right) \to 0$ as $\lambda \to 0$.

Furthermore, note that for $\lambda = 1$ the perspective function reduces to $g(v) \le 0$, and, for $\lambda = 0$, it reduces to $0 \le 0$.

In this way, substituting (10.5) and (10.6) for each disjunction $k \in K$, substituting c_k in the objective function, and converting the logic $\Omega(Y) = \text{True}$ into linear inequalities $A\lambda \le a$ (see Section 7.1), and by representing the Boolean variables Y_{jk} with the variables λ_{jk}, the NLP relaxation of problem (GDP) is given as the following continuous relaxation problem (CRP):

$$(\text{CRP}): \quad \min \ Z^L = \sum_{k \in K} \sum_{j \in J_k} \lambda_{jk} \gamma_{jk} + f(x)$$

$$\text{s.t.} \quad r(x) \le 0$$

$$x = \sum_{j \in J^k} v_{jk} \quad k \in K$$

$$\lambda_{jk} g\left(v_{jk} / \lambda_{jk}\right) \le 0 \quad j \in J_k, \quad k \in K \tag{10.7}$$

$$\sum_{j \in J_k} \lambda_{jk} = 1 \qquad k \in K$$

$$0 \le v_{jk} \le \lambda_{jk} U \qquad j \in J_k, \quad k \in K$$

$$A\lambda \le a$$

$$0 \le x \le U, v_{jk} \ge 0, \ \lambda_{jk} \ge 0 \quad j \in J_k, k \in K$$

where problem (CRP) in (10.7), which is a convex NLP, has the following property.

Property 10.2 The solution $\left(Z^L\right)^*$ of problem (CRP) yields a lower bound to solutions of GDP. This property readily follows from the fact that problem (CRP) in (10.7) is a continuous relaxation of problem (GDP).

We note that an MINLP reformulation of problem (GDP) is simply given by specifying in (CRP) the λ_{jk} variables to be binary, that is $\lambda_{jk} = 0, 1$, that is

$$(\text{MINLP-HR}) \quad \min \quad Z = \sum_{k \in K_j} \sum_{j \in J_k} \gamma_{jk} \lambda_{jk} + f(x)$$

$$\text{s.t.} \quad r(x) \leq 0$$

$$x = \sum_{j \in J^k} v_{jk} \quad k \in K$$

$$\lambda_{jk} g\left(v_{jk} / \lambda_{jk}\right) \leq 0 \quad j \in J_k, \quad k \in K \tag{10.8}$$

$$\sum_{j \in J_k} \lambda_{jk} = 1 \quad k \in K$$

$$0 \leq v_{jk} \leq \lambda_{jk} U \quad j \in J_k, \quad k \in K$$

$$A\lambda \leq a$$

$$0 \leq x \leq U, \quad v_{jk} \geq 0 \quad \lambda_{jk} = 0, 1 \quad j \in J_k, \quad k \in K.$$

This MINLP is denoted as the hull-relaxation formulation (MINLP-HR) since it consists of intersecting the convex hull of each disjunction (Balas, 1985; Ceria and Soares, 1999).

An alternate formulation to MINLP-HR is the big-M reformulation, in which the inequalities are replaced by big-M constraints, the c_k variables are substituted in the objective function, the logic $\Omega(Y) = \text{True}$ is converted into linear inequalities $A\lambda \leq a$, and the Boolean variables Y_{jk} are represented by the 0-1 variables λ_{jk}. The big-M formulation is then as follows:

$$(\text{MINLP-BM}) \quad \min \quad Z = \sum_{k \in K_j} \sum_{j \in J_k} \gamma_{jk} + f(x)$$

$$\text{s.t.} \quad r(x) \leq 0$$

$$g_{jk}(x) \leq M_{jk}\left(1 - \lambda_{jk}\right) j \in J_k, \quad k \in K$$

$$\sum_{j \in J_k} \lambda_{jk} = 1 \quad k \in K \tag{10.9}$$

$$A\lambda \leq a$$

$$0 \leq x \leq U, \quad \lambda_{jk} = 0, 1 \quad j \in J_k, \quad k \in K.$$

The relaxation of MINLP-BM, also a convex NLP, yields a lower bound to problem (GDP) that obeys the following property.

Property 10.3 The relaxation of MINLP-BM is weaker or equal to the relaxation of MINLP-HR. That is, their corresponding optimal objectives yield the following inequality:

$$Z_{\text{BM}}^R \leq Z_{\text{CRP}}. \tag{10.10}$$

This means that generally we can expect the hull relaxation to provide stronger lower bounds than the big-M formulation. However, we should note that the advantage of the latter is its smaller size compared to the hull-relaxation formulation. Which formulation will require less computational time depends on the relative strength of both bounds and their corresponding sizes. For computational experience see Lee and Grossmann (2000).

A difficulty in applying model MINLP-HR is that the perspective function $\lambda g(v/\lambda)$ is nondifferentiable at $\lambda = 0$, thus potentially causing failures in the solution of this problem. To overcome this problem, we replace the perspective function by the approximation proposed by Furman et al. (2020). In particular, we consider, for $\lambda g(v/\lambda) \leq 0, 0 \leq v \leq U\lambda$, the approximation inequality

$$[(1 - \varepsilon)\lambda + \varepsilon]g(v/[(1 - \varepsilon)\lambda + \varepsilon]) - \varepsilon g(0)(1 - \lambda) \leq 0. \tag{10.11}$$

We can verify that if we set $\lambda = 0$, which implies $0 \leq v \leq 0$, the inequality reduces to $\varepsilon g(0) - \varepsilon g(0) = 0 \leq 0$, and therefore it is exact.

On the other hand, if we set $\lambda = 1$, the inequality reduces to $g(v/1) - \varepsilon g(0)(0) = g(v) \leq 0$, which again is exact.

In addition to the desirable property that the approximation is exact at $\lambda = 0$ and $\lambda = 1$, the approximation function in (10.11) is convex in v and λ if $g(\cdot)$ is convex (see Exercise 10.2). Thus, this approximation has been used when implementing the hull reformulation.

10.3 Special Purpose Methods for GDP

10.3.1 Disjunctive Branch and Bound

First, we consider the disjunctive branch and bound method for solving problem (GDP). Rather than presenting detailed steps, we illustrate the main idea with Fig. 10.2, in which it can be seen that we start at the root node by solving the relaxation problem (CRP) in (10.7) (or alternatively the relaxation of the big-M formulation in (10.9)) in order to obtain the initial lower bound. Assuming a noninteger solution is obtained for any of the variables λ_{jk}, we typically branch on the disjunction k, which has a term j whose variable λ_{jk} is closest to one. We then create a new node by imposing the constraints of that corresponding term j of disjunction k and add to it the hull relaxation of the remaining disjunctions. For the complementary node, we take the convex hull of the remaining terms in the disjunction j and also add to it the hull relaxation of the remaining disjunction. Both nodes yield new lower bounds or an upper bound if a feasible solution is found to problem (GDP).

Figure 10.2 General branching rule for disjunctive branch and bound.

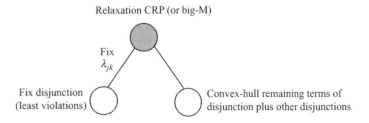

Relaxation CRP (or big-M)

Fix λ_{jk}

Fix disjunction (least violations)

Convex-hull remaining terms of disjunction plus other disjunctions

To illustrate the disjunctive branch and bound consider the following **GDP** problem (Lee and Grossmann, 2000):

$$\min \quad Z = (x_1 - 3)^2 + (x_2 - 2)^2 + c$$

$$\text{s.t.} \quad \begin{bmatrix} Y_1 \\ (x_1)^2 + (x_2)^2 - 1 \le 0 \\ c = 2 \end{bmatrix}$$

$$\vee \begin{bmatrix} Y_2 \\ (x_1 - 4)^2 + (x_2 - 1)^2 - 1 \le 0 \\ c = 1 \end{bmatrix} \tag{10.12}$$

$$\vee \begin{bmatrix} Y_3 \\ (x_1 - 2)^2 + (x_2 - 4)^2 - 1 \le 0 \\ c = 3 \end{bmatrix}$$

$$0 \le x_1, x_2 \le 8, \quad c \ge 0, \quad Y_j \in \{\text{True, False}\}\ j = 1, 2, 3.$$

Figure 10.3 shows the three subregions S_1, S_2, and S_3, corresponding to each term of the disjunction. As it can also be seen, the global optimum of this **GDP** problem of $Z^* = 1.172$, lies at the point $x_1 = 3.293$, $x_2 = 1.707$, which in turn lies inside the subregion S_2 from the second disjunctive term.

If we start the branch and bound search by solving the relaxed **GDP** problem (CRP) in (10.7), we obtain a lower bound of $Z^L = 1.154$ at the point $x_1 = 3.159$, $x_2 = 1.797$, which is infeasible for the **GDP** problem. However, as can be seen in Fig. 10.4 that point is closest to subregion S_2. That in fact is reflected by the values of the λ_j variables, namely $\lambda_1 = 0.016$, $\lambda_2 = 0.995$, $\lambda_3 = 0.029$, where it can be clearly seen that λ_2 is the closest to the value of one because it is in fact closest to the subregion S_2.

It is then logical to branch on subregion 2 by setting $\lambda_2 = 1$ on one node, while for the other node we evaluate the hull relaxation of the intersection of subregions S_1 and S_3. As can be seen in Fig. 10.5a, we obtain an upper bound $Z^U = 1.172$ for the first node. For the second node where we take the hull relaxation of S_1 and S_3, we obtain a lower bound

Figure 10.3 Geometric interpretation of GDP in (10.12).

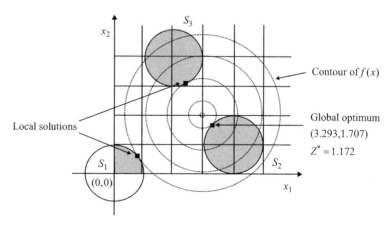

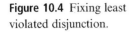

Figure 10.4 Fixing least violated disjunction.

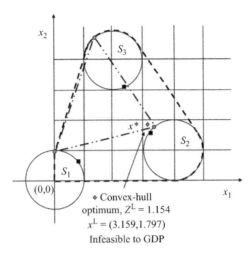

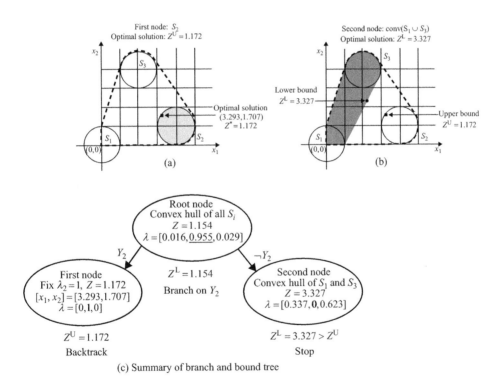

Figure 10.5 Branch and bound search applied to GDP (10.12).

$Z^L = 3.327$ (Fig. 10.5b) which is infeasible as it satifies neither of the constraints in subregions S_1 and S_3. Since that lower bound $Z^L = 3.327$ exceeds the value of the current upper bound $Z^U = 1.172$, this proves that the optimal GDP solution is given at that node in which $x_1 = 3.293$, $x_2 = 1.707$. The logic that has been used in this disjunctive branch and bound is summarized in Fig. 10.5c.

10.3.2 Logic-Based Outer Approximation

We describe in this section the logic-based outer-approximation algorithm by Türkay and Grossmann (1996). For simplicity, in the derivation of this algorithm we consider a special case of the problem (GDP), namely one that involves only two terms, and where the negation involves setting to zero a subset of the continuous variables. The special case is then given by the following GDP in (10.13):

$$(\text{GDP}') \quad \min \quad Z = \sum_{k \in K} c_k + f(x)$$

$$\text{s.t.} \quad r(x) \le 0$$

$$\begin{bmatrix} Y_k \\ g_k(x) \le 0 \\ c_k = \gamma_k \end{bmatrix} \vee \begin{bmatrix} \neg Y_k \\ B^k x = 0 \\ c_k = 0 \end{bmatrix} \quad k \in K \tag{10.13}$$

$$\Omega(Y) = \text{True}$$

$$0 \le x \le U, \quad c_k \in \mathbb{R}^1, \quad Y_k = \{\text{True}, \text{False}\},$$

where the B^k matrix is introduced for setting to zero a subset of x when $Y_k = \text{False}$, namely,

$$B_{ii} = \begin{cases} 1 & \text{if } x_i \text{ set to zero} \\ 0 & \text{if } x_i \text{ not set to zero} \end{cases}$$

$$B_{ij} = 0 \quad i \ne j.$$

Note that this structure arises in process-synthesis problems where the Boolean variables Y_k indicate whether a given unit k is selected or not.

The basic idea in the logic-based outer-approximation algorithm is to solve a sequence of NLP and GDP or MILP master problems in a similar manner to the original method. The big difference, however, is that a reduced NLP is solved, in which only the equations describing the selection of a given unit k are included as well as the definition of zero values for the units that are not selected. The reduced NLP (r-NLP) for fixed $\bar{Y}_k$ is then given as follows:

$$(r\text{-NLP}) \quad \min \quad Z = \sum_{k \in K} c_k + f(x)$$

$$\text{s.t.} \quad r(x) \le 0$$

$$\left. \begin{array}{l} g_k(x) \le 0 \\ c_k = \gamma_k \end{array} \right\} \text{for } \bar{Y}_k = \text{True} \tag{10.14}$$

$$\left. \begin{array}{l} B^k x = 0 \\ c_k = 0 \end{array} \right\} \text{for } \bar{Y}_k = \text{False}$$

$$0 \le x \le U.$$

In order to define the master problem, we assume we are given linearizations at iterations $\ell = 1, \ldots, L$ and iterations $L_k = \{\ell \mid Y_k = \text{True iteration } \ell\}$. Notice that this requires that a number of reduced NLPs (r-NLP) are solved so as to cover all the disjunctive terms that are true. For small or simple problems, these reduced NLP problems can be easily identified. As an example, consider the simple superstructure in Fig. 10.6 consisting of four units arranged

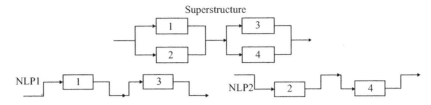

Figure 10.6 Example of superstructure optimization and two initial alternatives for generating linearizations through NLP subproblems NLP1 and NLP2.

in two processing stages, and where only one unit can be selected at each stage. A simple way to generate two reduced NLPs is to consider one alternative with units 1 and 3, and the other with units 2 and 4. Each of these alternatives gives rise to two reduced NLPs of the form (r-NLP), whose solutions can then be used to obtain linearizations for the four units.

For more general problems, one can formulate a set-covering problem (see Section 6.1.3.4) to identify the fewest reduced NLPs that cover all the true terms of the disjunction. The linear GDP master problem M_{GDP}^{OA} is then given as follows:

$$
M_{GDP}^{OA}: \quad \min \quad Z_{OA}^{L} = \alpha
$$

$$
\text{s.t.} \quad \left. \begin{array}{l} \alpha \geq \sum_{k \in K} c_k + f\left(x^{\ell}\right) + \nabla f\left(x^{\ell}\right)\left(x - x^{\ell}\right) \\[2mm] r\left(x^{\ell}\right) + \nabla r\left(x^{\ell}\right)^{T}\left(x - x^{\ell}\right) \leq 0 \end{array} \right\} \ell = 1, \ldots, L
$$

$$
\begin{bmatrix} Y_k \\ g_k\left(x^{\ell}\right)\nabla g\left(x^{\ell}\right)^{T}\left(x - x^{\ell}\right) \leq 0 \quad \ell \in L_k \\ c_k = \gamma_k \end{bmatrix} \vee \begin{bmatrix} \neg Y_k \\ B^k x = 0 \\ c_k = 0 \end{bmatrix} \quad k \in K \tag{10.15}
$$

$$
\Omega(Y) = \text{True}
$$

$$
0 \leq x \leq U, \quad c_k \in \mathbb{R}^1 \quad Y_k = \{\text{True}, \text{False}\}.
$$

This master problem can be reformulated as an MILP by using either the big-M formulation or the hull-relaxation formulation described in Chapter 7. The master problem can also be solved with a disjunctive branch and bound method as described in Section 10.3.1.

The steps of the logic-based OA algorithm are then as follows.

Step 0. Solve reduced NLPs (r-NLP) to cover all the disjunctions, i.e., $L_k \neq \emptyset \quad k \in K$.

Step 1. Solve M_{GDP}^{OA} with either of the two following solution approaches.

    (a) Transform to MILP hull relaxation or big-M $\rightarrow Z_{OA}^{L}$.
    (b) Use disjunctive branch and bound to determine Y^{new}.

Step 2. Solve NLP (r-NLP) for $Y^{new} \rightarrow Z^{U}$ upper bound.

Continue until $Z_{OA}^{L} \simeq Z^{U}$.

We should note that, for nonconvex GDP problems, one can add slack variables as in the MILP master (9.23) in Chapter 9 for the OA algorithm with augmented penalty and equality relaxation.

EXERCISES

10.1 Consider the case of disjunctions with linear constraints,

$$\bigvee_{j \in J_k} \left[A_{jk} x \le b_{jk} \right].$$

Show that the nonlinear convex hull given by (10.5) and (10.6) reduces to

$$x = \sum_j v_{jk} \qquad k \in K$$

$$A_{jk} v_{jk} \le b_{jk} \lambda_{jk} \qquad j \in J_k, \quad k \in K$$

$$\sum \lambda_{jk} = 1 \qquad k \in K$$

$$0 \le v_{jk} \le U \lambda_{jk} \qquad j \in J_k, \quad k \in K.$$

10.2 Prove that, for the perspective function $\lambda g(v/\lambda) \le 0$, and for which $0 \le v \le U\lambda$, the left-hand side in the approximation inequality is convex if $g(\cdot)$ is convex:

$$[(1 - \varepsilon)\lambda + \varepsilon] g(v/[(1 - \varepsilon)\lambda + \varepsilon]) - \varepsilon g(0)(1 - \lambda) \le 0.$$

10.3 Given the figure below, formulate the MINLP problem in which either x_1 and x_2 lie within the area of the circle with a cost $2x_1 + x_2$, or else lie at the origin with zero cost.
(a) Formulate the problem first as a nonlinear GDP.
(b) Transform it into an MINLP without using big-M constraints and simplify it by eliminating disaggregated variables.

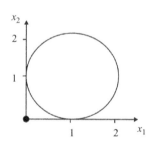

10.4 For the generalized disjunctive program given below, do the following.
(a) Reformulate it as an MINLP using the convex-hull formulation for the disjunction.
(b) Reformulate it as a big-M MINLP ($M = 50$).
(c) Solve both reformulations and compare their relaxations.

$$\min Z = c + (x_1 - 3)^2 + (x_2 - 2)^2$$

$$st$$

$$\begin{bmatrix} Y_1 \\ x_1^2 + x_2^2 \le 1 \\ c = 2 \end{bmatrix} \vee \begin{bmatrix} Y_2 \\ (x_1 - 4)^2 + (x_2 - 1)^2 \le 1 \\ c = 1 \end{bmatrix} \vee \begin{bmatrix} Y_3 \\ (x_1 - 2)^2 + (x_2 - 4)^2 \le 1 \\ c = 3 \end{bmatrix}$$

$$0 \le x_1 \le 8, 0 \le x_2 \le 8, \quad Y_j = True, False, j = 1, 2, 3$$

11 Constraint Programming

11.1 Logic-Based Modeling

Constraint programming (CP) is a logic-based approach to optimization that makes use of implicit functions and logic inference (Van Hentenryck, 1989; Hooker, 2002; Hooker and van Hoeve, 2018). There is no canonical form in constraint programming as it provides flexibility for a variety of variables and constraints that can be handled. Specifically, we have the following.

(a) Variables: continuous, Boolean, discrete.
(b) Constraints:
- algebraic $h(x) \leq 0$;
- disjunctions $[A_1 x \leq b_1] \vee [A_2 x \leq b_2]$;
- conditional constraints
   if $x = y \Rightarrow zy \leq w$;
- global constraints, for example,
   all different $(x_i, i = 1, \ldots, n)$ x_i-integer;
- cumulative constraint
   Cum. $((t_1, t_2, \ldots, t_n), (D_1, D_2, \ldots, D_n), c_1, c_2, \ldots, c_n), C$
   t_i start time, job D_i consumes resource c_i, C resource limit;
- meta constraints

$$(x \neq y) + (y \neq z) + (z \neq x) \geq 2.$$

Constraint programming is largely aimed at finding feasible solutions to a set of constraints such as those shown above, but it is also possible to optimize for a given objective function. However, as we will see below, the search will consist of finding a sequence of feasible solutions with a specified decrease in the objective function.

An example of a CP model is the following:

$$
\begin{aligned}
&\text{Find } x, y, z \\
&\quad \text{Continuous: } x \geq -5 \quad y \geq 0 \\
&\quad \text{Integer: } z = 0, 1, 2, 3, 4 \\
&\quad x^3 + 10x = y^x - 2^z \\
&\quad zx + 7.7y = 2.4 \\
&\quad (z - 1)^{y+1} \leq 10 \\
&\quad \{[\ell n(y + 2x + 12) \leq z + 5] \vee [y \geq z^2]\} \Rightarrow \{x \leq 0 \wedge y \leq 1\}\} \\
&\quad x \leq 0 \Rightarrow z > 3.
\end{aligned}
\tag{11.1}
$$

The solution to the CP problem in (11.1) is $x = -1.28505$, $y = 0.979$, $z = 4$.

It is interesting to note that the CP problem in (11.1) can be reformulated as a GDP problem by reformulating the last two constraints by using the following equivalent logic statements (see Chapter 7, Section 7.1).

For the second to last $A \vee B \Rightarrow C \wedge D$ is equivalent to $\neg A \wedge \neg B \vee (C \wedge D)$.

For the last constraint $C \Rightarrow E$ is equivalent to $\neg C \vee E$.

By rewriting the last two logic expressions in (11.1) using these equivalent statements, the CP in (11.1) can be rewritten as the following GDP problem in which we specify a zero value for the objective as it is a feasibility problem:

$$\min \quad Z = 0$$

$$\text{s.t.} \quad x^3 + 10x = y^x - 2^z$$

$$zx + 7.7y = 2.4$$

$$(z - 1)^{y+1} \leq 10$$

$$\begin{bmatrix} Y_1 \\ \ell n(y + 2x + 12) - z - 5 \leq \varepsilon \\ y - z^2 \leq -\varepsilon \end{bmatrix} \vee \begin{bmatrix} Y_2 \\ x \leq 0 \\ y \leq 1 \end{bmatrix} \tag{11.2}$$

$$\begin{bmatrix} W_1 \\ x \geq \varepsilon \end{bmatrix} \vee \begin{bmatrix} W_2 \\ z \geq 4 \end{bmatrix}$$

$$x \geq -5, y \geq 0, z = 0, 1, 2, 3, 4, \ Y_1, W_1 = \text{True}, \ Y_2, W_2 = \text{False}.$$

As a second example, we consider the traveling salesman problem that we considered in Chapter 6. First, assume we label all the cities with an integer $n = 1, 2, \ldots, N$. Let y_k be an integer variable that represents city n visited in kth position of the tour. For instance, if $y_1 = 3$, city 3 is visited first. Furthermore, let $c_{y_k y_{k+1}}$ be the travel cost from city y_k to y_{k+1}. Notice that we are defining a cost coefficient that has as indices the variables y_k and y_{k+1}. In this way, the TSP problem can be formulated as the following CP problem

$$\text{Select } y_k \quad 1, \ldots, N$$

$$\min \sum_{k=1}^{N} c_{y_k y_{k+1}} \tag{11.3}$$

$$\text{s.t. all different } (y_k, k = 1, \ldots, N),$$

where the all-different constraints ensure that different integer labels of the cities are assigned to the variables y_k. Notice that, with (11.3), we can write the traveling salesman problem in a very compact form given the capability of handling indices as variables and given the implicit constraint "all different."

11.2 Search in Constraint Programming

11.2.1 Domain Reduction and Constraint Propagation

The search for a feasible and/or optimum solution in CP relies on an implicit enumeration of a tree in which no LP relaxation is solved, but instead a domain reduction and constraint propagation are performed at each node.

The first major element of the CP search is a depth-first search on the tree of discrete or interval alternatives. Lower bounds are simply obtained by partial solutions in the tree, while upper bounds correspond to feasible solutions found in the tree enumeration.

The second major element of the CP search is domain reduction and constraint propagation, which in many ways is unique to CP, and often a major reason for its successful performance. Domain reduction and constraint propagation do not rely on general-purpose algorithms, because they correspond to specific algorithms for different classes of constraints.

An example is bounds tightening for linear constraints. Consider the constraints,

$$\sum_j c_j x_j \in [L, U]$$
$$x_j \in [\ell_j, u_j],$$

(11.4)

where $c_j, L, U > 0$.

Because $\sum_j c_j x_j \geq L \Rightarrow x_i \geq L/c_i - \sum_{j \neq i} c_j x_j / c_i$, this in turn means that the smallest lower bound occurs at the upper bound, $x_j = u_j$, and hence, the tightest lower bound of x_i is given by the inequality,

$$x_i \geq L/c_i - \sum_{j \neq i} c_j u_j / c_j \quad \forall i.$$

(11.5)

Similarly, because $\sum_j c_j x_j \leq U \Rightarrow x_i \geq U/c_i - \sum_{j \neq i} c_j x_j / c_i$, this in turn means that the largest upper bound occurs at the lower bound, $x_j = l_j$, and hence the tightest upper bound of x_i is given by the inequality

$$x_i \leq U/c_i - \sum_{j \neq i} c_j l_j / c_i \quad \forall i.$$

(11.6)

In summary, we can set as a result of the tightening of bounds the following:

$$\text{Lower bound } x_i = \max \left\{ l_i, L/c_i - \frac{\sum_{j \neq i} c_j u_j}{c_i} \right\}$$

(11.7)

$$\text{Upper bound } x_i = \min \left\{ u_i, U/c_i - \frac{\sum_{j \neq i} c_j l_j}{c_i} \right\}.$$

(11.8)

(a) As a specific example, consider the linear equality with corresponding variable bounds,

$$2x + 3y = 11$$
$$x \geq 0 \quad y \geq 0. \tag{11.9}$$

For the case where the variables x, y are continuous, assume $L = 0, U = 11$. Then it follows from (11.8) that the tightest upper bound is given by

$$x = \frac{11 - 3y}{2}, y = 0 \Rightarrow x = 5.5$$

$$y = \frac{11 - 2x}{3}, x = 0 \Rightarrow y = 3.66.$$

Hence, we can conclude that the tightest bounds for the linear inequality in (11.9) are given by $x \in [0, 5.5] \quad y \in [0, 3.66]$.

(b) Consider now the case where x, y are integer in the linear inequality in (11.9). If we simply round down the fractional values obtained, the tightest bounds are given by $x \in [0, 5], \quad y \in [0, 3]$.

We should note that, for nonlinear constraints, it is generally more difficult to infer tight bounds unless the constraints have the property of monotonicity.

One additional example for domain reduction and constraint propagation is "edge-finding conditions" for job-shop scheduling. Consider a pair of jobs with corresponding start times and duration:

job i Ts_i start time job i duration d_i
job j Ts_j start time job j duration d_j

and for which the following disjunction holds:

$$\left[Ts_i + d_i \leq Ts_j\right] \vee \left[Ts_j + d_j \leq Ts_i\right]. \tag{11.10}$$

That is, job i is executed before job j, or job j is executed before job i. The disjunction in (11.10) can be expressed in terms of 0-1 and continuous variables as was shown in Chapter 7 (Section 7.2). Instead, in CP the disjunction is resolved by a procedure that is based on analyzing earliest starts and latest ends. The procedure is known as "edge-finding conditions" and was developed by Applegate and Cook (1991).

Let est = earliest start, and lte = latest end.

If $\left(lte_i - est_i \leq d_j + d_i\right)$ and $\left(lte_j - est_i \leq d_j\right)$ then job j must be executed before job i. Consider the example in Fig. 11.1.

Applying the edge-finding conditions to the data in Fig. 11.1, this implies that job j must take place before job i.

11.2.2 Tree Search

The search in CP consists of a depth-first enumeration of a tree with domain reduction at each node (or logic inference; arc consistency).

Figure 11.1 Jobs i and j with corresponding earliest start times (est) and latest end times (lte).

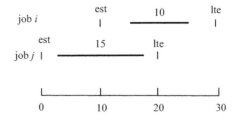

Figure 11.2 Tree enumeration in CP with constraint propagation.

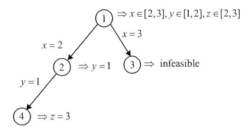

To illustrate the tree enumeration in CP, consider the following example:

$$\text{Find integer } x, y, z$$
$$x \in [1, 3] \; y \in [1, 3] \; z \in [1, 3]$$
$$\text{s.t.} \quad y < z \tag{11.11}$$
$$x - y = 1$$
$$x \neq z.$$

Notice that the strict inequality $y < z$ can be specified in CP, whereas in mathematical programming no strict inequalities are allowed. But because the variables are integers one would express that constraint in mathematical programming as $y \leq z - 1$.

It can be shown that the maximum potential number of nodes in the tree search is 40. By using the "generate and test" paradigm and by using domain reduction and constraint propagation the search requires many fewer nodes and is as follows (see Fig. 11.2).

At the initial node 1, we can infer from $x - y = 1$, that $x \in [2, 3]$ and $y \in [1, 2]$, which in turn means that, for $x = y + 1$ and $y \leq z - 1$ or $y + 1 \leq z$, this implies $z \in [2, 3]$.

If we start branching with x, we have $x = 2$, from $x - y = 1$, this implies $y = 1$ (node 2). On the other hand, if we set $x = 3$, because $x - y = 1$, this implies $y = 2$. Furthermore, $y + 1 \leq z$, which implies $z = 3$. But this implies that the constraint $x \neq z$ is violated. Hence, node 3 is infeasible and can be fathomed.

If we now branch on node 4, by setting $y = 1$, we have that because $x \neq z$ must be satisfied, $z = 1$ or $z = 3$. Since $y + 1 \leq z$, this implies we can set $z = 3$ at node 4. As no more nodes are open, this proves that $x = 2$, $y = 1$, $z = 3$ is the feasible solution satisfying the inequalities in (11.11).

This simple example, then, illustrates the capability of inference and constraint propagation of CP. Note that in "classical" branch and bound, the generate and test method ends up typically enumerating a larger number of nodes because it usually does not perform domain reduction and constraint propagation (except for recent versions of CPLEX, GUROBI, and XPRESS).

In summary, CP problems are represented by a compact syntax in specialized modeling software (e.g., OPL, Van Hentenryck et al., 1989). Although CP is aimed at feasibility/discrete problems, optimization can be performed by solving a sequence of feasibility problems, where we update the upper bound $f(x) \leq UB - \delta$, where δ is the specified decrease. The optimization in CP performs reasonably well on simple objective functions (e.g., makespan in scheduling), but does not perform as well if there are many variables in the objective function (e.g., total cost).

EXERCISES

11.1 The all-different constraints $y_i \neq y_j \quad i \neq j$, where y_i are integer variables n, can be regarded as a disjunction

$$\vee_{k=1,K} \begin{bmatrix} z_{ik} \\ y_i = k \end{bmatrix} \quad \forall i$$

in which, for z_{ik} = True, this implies that $y_i = k$.

Show that by using the convex full formulation of this disjunction, the "all-different constraint" can be modeled through the following set of 0-1 linear constraints,

$$y_i = \sum_k k z_{ik} \quad \forall i$$

$$\sum_k k z_{ik} = 1 \quad \forall i$$

$$\sum_i z_{ik} \leq 1 \quad \forall k$$

$$z_{ik} = 0, 1.$$

11.2 Verify in Fig. 11.1 that job j must be executed before job i, both by inspection and by applying the "edge-finding" conditions.

12 Nonconvex Optimization

12.1 Major Approaches to Global Optimization

When addressing the solution of NLP problems (12.1) or more generally MINLP problems (12.2) these give rise to nonconvex optimization problems (Horst and Tuy, 1996; Tawarlamani and Sahinidis, 2004). In the case of NLP problems, the nonlinear equations $h(x) = 0$ generally introduce nonconvexities unless they can be shown to relax as convex inequalities. Furthermore, if either the objective function $f(x)$ is nonconvex and/or the inequalities $g(x) \leq 0$ are nonconvex, the NLP will be nonconvex which means that there is a distinct possibility that it may exhibit more than one local minima:

$$
\begin{aligned}
\min \quad & f(x) \\
\text{s.t.} \quad & h(x) = 0 \\
& g(x) \leq 0 \\
& x \in R^n.
\end{aligned}
\tag{12.1}
$$

In the case of MINLP problems, the nonconvexity first arises from the fact that the discrete variables y define a nonconvex feasible region. This, in turn, implies that MILP problems, despite having linear objective and constraints, are also nonconvex. Similarly, as in NLP problems, nonconvexities may be due to the nonlinear equations and nonconvex objective function and inequalities:

$$
\begin{aligned}
\min \quad & f(x, y) \\
\text{s.t.} \quad & h(x, y) = 0 \\
& g(x, y) \leq 0 \\
& x \in X, \quad y \in Y.
\end{aligned}
\tag{12.2}
$$

The three major approaches that are used in global optimization are: (a) convexification, (b) heuristics, and (c) global optimization. The approach based on heuristics does not yield rigorous global-optimum solutions and relies on computational strategies such as multistart strategies from different initial points, typically generated at random. It can also rely on modifying subproblems such as the MILP master problem of the OA algorithm in Section 9.4 of Chapter 9 in which slack variables and a penalty function are introduced to avoid cutting off a feasible solution (see (9.23)), and modifying termination criteria (e.g., continue search until no improvement is found). We should also note that, for the particular case of concave separable objective and linear constraints, the problem can be approximated through MILP using special ordered sets, as described in Beale and Forrest (1976).

We discuss in the next sections the approaches based on convexification and global optimization.

12.2 Convexification

In a number of special cases it is possible to convexify an NLP problem or the NLP relaxation of an MINLP problem so that the corresponding problems yield a unique local optimum solution.

The presence of posynomials is a particular structure that can be exploited for convexifications. A posynomial is formally defined as a function of the following form, consisting of sums of products of variables raised to an arbitrary exponent,

$$f(x) = \sum_i c_i \prod_j x_j^{\alpha_{ij}}, \tag{12.3}$$

where $c_i > 0$ and α_{ij} is an arbitrary exponent. These functions arise in a special class of NLP problems known as geometric programs (Duffin et al., 1967), in which the objective function and constraints correspond to posynomials.

Consider the transformation

$$x_j = e^{u_j}; \tag{12.4}$$

if we apply it to the posynomial in (12.3) this yields the transformed function,

$$f(u) = \sum_i c_i \prod_j e^{\alpha_{ij} u_j} = \sum_i c_i e^{\sum_j \alpha_{ij} u_j}, \tag{12.5}$$

which can easily be shown to be convex as it consists of a sum of positive convex functions.

As one specific example, consider the nonconvex inequality (linear fractional plus concave separable)

$$\frac{x_1}{x_2} + x_1^{0.5} \leq 1. \tag{12.6}$$

Substituting (12.4) yields

$$e^{u_1} e^{-u_2} + e^{0.5 u_1} \leq 1. \tag{12.7}$$

This, in turn, can be rearranged as

$$e^{u_1 - u_2} + e^{0.5 u_1} \leq 1, \tag{12.8}$$

which corresponds to a convex inequality since both terms are convex.

Special cases are monomials, in which we have an inequality with a posynomial at each side of the inequality. As an example, consider the monomial

$$\frac{x_1}{x_2} \leq x_1^{0.5}. \tag{12.9}$$

This implies, by substituting (12.4),

$$e^{u_1} e^{-u_2} \leq e^{0.5u_1} \tag{12.10}$$

or

$$e^{u_1 - u_2} \leq e^{0.5u_1}. \tag{12.11}$$

And finally, by taking natural logs on both sides, this yields

$$u_1 - u_2 \leq 0.5u_1, \tag{12.12}$$

which is a linear inequality. Thus, monomials can always be transformed into linear inequalities.

12.3 Global Optimization of Bilinear Programs

Global-optimization methods can be classified first as deterministic methods in which global solutions can be mathematically guaranteed for structured problems that include the following types of functions (Tawarlamani and Sahinidis, 2004):

– concave functions
– bilinear, linear fractional
– difference of convex functions.

These methods can guarantee global optimality because they rely on valid convex underestimators. As an example, Fig. 12.1 illustrates the use of a secant for underestimating concave functions. The secant is denoted as a convex envelope since it is the tightest underestimator.

  Global-optimization methods can also be classified as stochastic methods that can handle general types of functions. These methods include (Eiben and Smith, 2003):

Figure 12.1 Secant for underestimating concave functions.

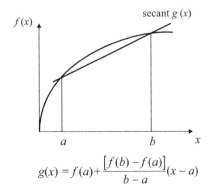

$$g(x) = f(a) + \frac{[f(b) - f(a)]}{b - a}(x - a)$$

– simulated annealing
– genetic algorithms.

These methods unfortunately have no mathematical guarantees of global optimality unless they are evaluated at an infinite number of trial points.

In this section, we focus on deterministic global-optimization methods for bilinear programs (Sherali and Alameddine, 1992; Quesada and Grossmann, 1995) in order to illustrate the main ideas behind deterministic global-optimization methods. The general form of a bilinear NLP is as follows:

$$\text{NLP-}b: \quad \min \ Z = f^0(x)$$
$$\text{s.t.} \quad f^\ell(x) \leq 0 \quad \ell = 1, 2, \ldots, L$$
$$x \in X, \quad x^L \leq x \leq x^U,$$

where $X = \{x | x \in \mathbb{R}^n, Ax \leq b\}$ and $f^\ell = \sum_{i \in I} \sum_{j \in J} c_{ij}^\ell x_i x_j + g^\ell x, \ \ell = 0, 1, \ldots, L,$ where $g^\ell(x)$ are convex functions.

The most common method for solving global-optimization problems such as the bilinear NLP-b, is the spatial branch and bound method (Horst and Tuy, 1996). This method relies on three basic elements:

– LP or NLP relaxation that yields lower bounds,
– upper bounds given by a local or feasible solution to nonconvex
  NL-b,
– search in the space of continuous variables by generating subregions that are to be analyzed by a branch and bound search.

The LP relaxation for bilinear terms is given by the McCormick (1976) convex envelopes. These are derived as follows. Let $x, y,$ be two continuous variables specified within lower and upper bounds, $x^L \leq x \leq x^U, \ y^L \leq y \leq y^U$. We will use the principle that for $g_1 \geq 0, \ g_2 \geq 0 \Rightarrow g_1 * g_2 \geq 0$. Defining $g_1 = x^L - x \geq 0, \ g_2 = y^L - y \geq 0$, we then have

$$xy \geq xy^L + x^L y - x^L y^L. \tag{12.13}$$

Similarly, let $g_1 = x^U - x \geq 0, \ g_2 = y^U - y \geq 0,$ which implies $g_1 * g_2 = x^U y^U - x^U y - xy^U + xy \geq 0,$ or

$$xy \geq x^U y + xy^U - x^U y^U. \tag{12.14}$$

Notice that the right-hand side in the inequalities in (12.13) and (12.14) represent linear underestimators of the bilinear term xy.

Also, from $g_1 = x - x^L \geq 0, \ g_2 = y^U - y \geq 0, \ g_1 * g_2 = xy^U - xy - x^L y^U + x^L y \geq 0,$ which leads to

$$xy \leq xy^U + x^L y - x^L y^U, \tag{12.15}$$

and with $g_1 = x^U - x \geq 0, \ g_2 = y - y^L \geq 0, \ g_1 * g_2 = x^U y - x^U y^L - xy + xy^L \geq 0,$ or

Figure 12.2 Geometrical interpretation of McCormick convex envelopes for bilinear term xy.

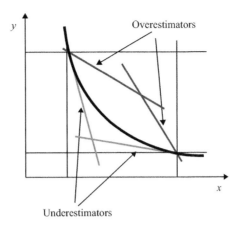

$$xy \leq xy^{\mathrm{L}} + x^{\mathrm{U}}y - x^{\mathrm{U}}y^{\mathrm{L}}. \tag{12.16}$$

Notice that (12.15) and (12.16) represent linear overestimators of the bilinear term xy. In summary, the inequalities (12.13)–(12.16) represent the McCormick convex envelopes of the bilinear term xy. These convex envelopes, which correspond to the tightest underestimators and overestimators of the bilinear term xy, are displayed in Fig. 12.2.

The following property holds for the McCormick envelopes.

Property 12.1 The inequalities (12.13)–(12.16) are exact at the bounds of xy (boundary of xy).

As an illustration, from (12.13) set $x = x^{\mathrm{L}}$. This then leads to $x^{\mathrm{L}}y \geq x^{\mathrm{L}}y^{\mathrm{L}} + x^{\mathrm{L}}y - x^{\mathrm{L}}y^{\mathrm{L}} = x^{\mathrm{L}}y^{\mathrm{L}}$, or $x^{\mathrm{L}}y \geq x^{\mathrm{L}}y^{\mathrm{L}}$, i.e., we obtain an exact approximation.

Making use of the convex envelopes in (12.13)–(12.16), we can define the relaxation problem RP for the bilinear problem NLP-b by substituting the inequalities for the corresponding bilinear terms $u_{ij} = x_i x_j$:

$$\min \; Z^R = \sum_{i \in I} \sum_{j \in J} c_{ij}^0 u_{ij} + g^0(x)$$

$$\text{s.t.} \; \sum_{i \in I} \sum_{j \in J} c_{ij}^\ell u_{ij} + g^\ell(x) \leq 0 \quad \ell = 1, \ldots, L$$

$$\left. \begin{aligned} u_{ij} &\geq x_i x_j^{\mathrm{L}} + x_i^{\mathrm{L}} x_j - x_i^{\mathrm{L}} x_j^{\mathrm{L}} \\ u_{ij} &\geq x_i x_j^{\mathrm{U}} + x_i^{\mathrm{U}} x_j - x_i^{\mathrm{U}} x_j^{\mathrm{U}} \\ u_{ij} &\leq x_i x_j^{\mathrm{U}} + x_i^{\mathrm{L}} x_j - x_i^{\mathrm{L}} x_j^{\mathrm{U}} \\ u_{ij} &\leq x_i x_j^{\mathrm{L}} + x_i^{\mathrm{U}} x_j - x_i^{\mathrm{U}} x_j^{\mathrm{L}} \end{aligned} \right\} i \in I, j \in J \tag{RP}$$

$$x^{\mathrm{L}} \leq x \leq x^{\mathrm{U}}, u^{\mathrm{L}} \leq u \leq u^{\mathrm{U}}, x \in X.$$

We can establish the following properties.

Property 12.2 The relaxation problem (RP) is a convex NLP problem.

This simply follows from the fact that problem (RP) involves a convex objective function and convex inequalities. Note that if $g^{\ell}(x)$ is linear, (RP) reduces to an LP problem. Furthermore, being convex, problem (RP) has a unique optimum solution.

Property 12.3 The relaxation problem (RP) yields a lower bound Z^R to problem (NLP-b); i.e., $Z^R \leq Z^*$.

This follows from the fact that we relaxed $u_{ij} = x_i y_j$ by the inequalities in (RP).

We should also note that not all convex envelopes may be required. For example, assume that a given bilinear term is only present in the objective function. In that case,

– if $c_{ij}^0 > 0$ we only need the lower bounding inequalities $u_{ij} \geq \cdots$ (12.13), (12.14);
– if $c_{ij}^0 < 0$ we only need the upper bounding inequalities $u_{ij} \leq \cdots$ (12.15), (12.16).

It is also important to note that tight bounds, x^L, x^U, are required to obtain a tight lower bound from the relaxation problem (RP). This can be accomplished by a bound-tightening procedure that involves successively minimizing and maximizing the value of each variable x_i (Zamora and Grossmann, 1999):

$$x^L \rightarrow \min x_i \qquad x^U \rightarrow \max x_i$$
$$\text{s.t. constr.(RP)} \qquad \text{s.t. constr.(RP).} \qquad (12.17)$$

The upper bound for the objective function can generally be obtained by finding a local solution to NLP-b. For the special case that the nonlinear terms are convex in NLP-b, a feasible point in that problem also yields an upper bound.

Finally, the branch and bound search is conducted in a similar way as for MILP and MINLP problems, except that the branching is performed through region elimination. In particular, if in a given subregion, when solving the corresponding relaxation problem (RP), the lower bound lies below the upper bound, then we partition the subregion; otherwise we reject the subregion. This partition is illustrated in Figs. 12.3 and 12.5 for one and two variable problems. In Fig. 12.4 we show the corresponding tree search after performing two partitions in the one variable problem of Fig. 12.3.

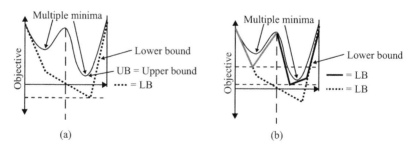

Figure 12.3 Successive partition of region in a one-dimensional problem.

Figure 12.4 Branch and bound tree from partitions in Fig. 12.3.

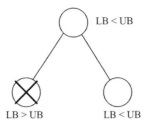

LB < UB

LB > UB   LB < UB

Figure 12.5 Successive elimination of subregions in two dimensions.

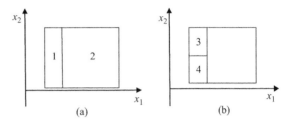

(a)   (b)

We should note that the variable bounds are updated in each subregion. Also, for the general case, we need to decide on what bilinear terms $x_i x_j$ to branch. A common branching rule is as follows.

(a) Select term $x_i x_j$ with largest difference from u_{ij}, $d_{ij} = |u_{ij} - x_i x_j|$.
(b) Determine for x_i the sum difference $\sum_{k \neq i} d_{ij}$ in all terms $x_i x_k$, $k \neq i$, and for x_j the sum difference $\sum_{k \neq j} d_{ij}$ in all terms $x_k x_j$, $k \neq j$.
(c) Select x_i or x_j with largest sum differences.

Qualitatively, the idea behind this branching rule is to select the variable that leads to the largest difference with respect to the variable u_{ij}.

Finally, as an example, consider the following bilinear problem, P.

$$P: \quad \min \quad f = -x - y$$
$$\text{s.t.} \quad xy \leq 4$$
$$0 \leq x \leq 6 \tag{12.18}$$
$$0 \leq y \leq 4.$$

The plot of the feasible region of problem P is shown in Fig. 12.6.
The relaxation problem (RP) is given as follows:

$$RP: \quad \min \quad f = -x - y$$
$$\text{s.t.} \quad u \leq 4$$
$$u \geq xy^L + x^L y - x^L y^L$$
$$u \geq xy^U + x^U y - x^U y^U$$
$$u \leq xy^L + x^U y - x^U y^L \tag{12.19}$$
$$u \leq xy^U + x^L y - x^L y^U$$
$$0 \leq x \leq 6 \quad\quad 0 \leq y \leq 4.$$

Figure 12.6 Plot of feasible region for problem P.

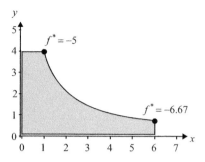

Figure 12.7 Plot of relaxation (RP).

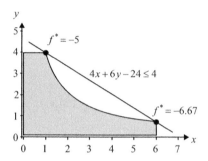

Substituting the corresponding bounds yields the LP problem,

$$
\begin{aligned}
\text{LP}: \quad \min \quad & f = -x - y \\
\text{s.t.} \quad & u \leq 4 \\
& u \geq 0 \\
& u \geq 4x + 6y - 24 \\
& u \leq 6y \\
& u \leq 4x \\
& 0 \leq x \leq 6 \quad 0 \leq y \leq 4.
\end{aligned}
\tag{12.20}
$$

This yields the solution $x = 6$, $y = 0.666$, $u = 4$, with objective value $f^* = -6.67$, which corresponds to the global-optimum solution. That is, the global optimum is obtained at the initial relaxation of problem NLP-b shown in Fig. 12.7. This is due to the fact that, because $u = 4$, the inequality $4x + 6y - 24 \leq 4$, which reduces to $4x + 6y \leq 28$, corresponds to the inequality shown in Fig. 12.7, which together with the bounds on the variables, defines the linear convex hull of the feasible region, which has a minimum value of -6.66 at the extreme point $(6, 0.666)$.

12.4   Global Optimization of More General Functions

We should note that the area of global optimization has expanded considerably beyond the solution of bilinear NLP problems. The trend has been to accommodate a large class of algebraic functions for which convex envelopes are developed for their use in spatial branch and bound methods (e.g., Quesada and Grossmann, 1995; Adjiman et al., 1998; Tawarmalani and Sahinidis, 2005). Global optimization has also been extended to MINLP problems (e.g., Kesavan et al., 2004; Misener and Floudas, 2013). Recent reviews on global-optimization algorithms can be found in Floudas and Gounaris (2009) and Tawarmalani and Sahinidis (2013). Appendix A summarizes the software such as BARON, Antigone, and SCIP, that is currently available for solving nonconvex NLP and MINLP problems.

EXERCISES

12.1 Replace the following nonlinear inequality by a linear inequality that rigorously bounds the feasible space defined by that inequality:

$x_1 x_2 \leq 8$, where $0 \leq x_1 \leq 4$, $0 \leq x_2 \leq 6$.

12.2 Determine whether the following inequalities can be reformulated as convex inequalities:

$$x_1 x_2 \leq 9$$
$$x_1 - 2x_2 \leq 0$$
$$x_1, x_2 \geq 1.$$

12.3 Assume that the following MINLP problem is solved with both the NLP-based branch and bound and the outer-approximation method:

$$\min \quad f_0(x) + a_0^T y$$
$$\text{s.t.} \quad f_j(x) + a_j^T y \leq b \quad j = 1, 2, \ldots, n$$
$$x^L \leq x \leq x^U, \ x \in R^n, \ y = \{0, 1\}^m,$$

where $f_j(x) = \sum_i c_i \prod_k x_k^{\alpha_{ik}}$, $c_i \geq 0$, and α_{ik} is arbitrary. Determine if each of these methods is guaranteed to find the global solution to this MINLP.

12.4 Consider the product $y*x$ in a constraint of an MINLP, where $y = 0, 1$. In order to linearize this term, it is proposed to use the Glover inequalities that involve a new variable $w = y*x$:

$$x - (1 - y)x^U - w \leq 0$$
$$w - x + (1 - y)x^L \leq 0$$
$$yx^L - w \leq 0$$
$$w - yx^U \leq 0.$$

Develop alternative constraints using disjunctive programming as a basis. Determine if your new constraints have a weaker, stronger or equivalent relaxation to Glover's inequalities.

12.5 Given the bilinear NLP below, find the global-optimal solution using the McCormick convex envelopes and a spatial branch and bound. To obtain good initial lower and upper bounds solve LPs for the bounds of the four continuous variables:

$$\min \quad f = x_1 - x_2 - y_1 - x_1 y_1 + x_1 y_2 + x_2 y_1 - x_2 y_2$$
$$\text{s.t.} \quad x_1 + 2x_2 \leq 8$$
$$4x_1 + x_2 \leq 12$$
$$3x_1 + 4x_2 \leq 12$$
$$2y_1 + y_2 \leq 8$$
$$y_1 + y_2 \leq 5$$
$$0 \leq x_1, x_2, y_1, y_2 \leq 10.$$

Verify your answer with the software packages BARON and ANTIGONE in GAMS. (Use OPTION NLP=BARON or ANTIGONE.)

13 Lagrangean Decomposition

13.1 Overview of Decomposition for Large-Scale Problems

When addressing the solution of large-scale optimization problems, it becomes necessary at some point to decompose them in order to avoid the solution of very large-scale problems, and thereby achieve reasonable computational times. Two major structures that arise in decomposition, and which we illustrate here for LP problems, are the following.

(a) LP problems with complicating variables:

$$
\begin{aligned}
\max \quad & a^T y + \sum_{i=1,\ldots,n} c_i^T x_i \\
\text{s.t.} \quad & Ay + D_i x_i = d_i && i = 1,\ldots,n \\
& y \geq 0, x_i \geq 0, && i = 1,\ldots,n.
\end{aligned}
\tag{13.1}
$$

The general form is as presented in the LP in (13.1), where y are complicating variables in the sense that if we fix values of the variables y, the problem decomposes into n subproblems in terms of x_i, $i = 1,\ldots,n$ variables. See also Fig. 13.1. This structure commonly arises in two-stage stochastic programming problems, and the corresponding method that is used is Benders decomposition (see Chapter 14). The problem also arises in multiperiod design problems where the y variables represent design decisions that must be selected for the n different periods of operation.

(b) LP problems with complicating constraints:

$$
\begin{aligned}
\max \quad & c^T x \\
\text{s.t.} \quad & Ax = b \\
& D_i x_i = d_i && i = 1,\ldots,n \\
& x \in X = \{x | x_i, i = 1,\ldots,n |, x_i \geq 0\}.
\end{aligned}
\tag{13.2}
$$

The general form is as presented in the LP in (13.2), where there are complicating constraints in the sense that if we remove the equations $Ax = b$, the problem decomposes into n subproblems in terms of the x_i, $i = 1,\ldots,n$ variables. See also Fig. 13.2. This structure commonly arises in multistage stochastic programming problems, and the corresponding method that is used is Lagrangean decomposition. The problem also arises in multiperiod planning problems where the complicating constraints represent inventory equations that interconnect successive time periods. We should note that it is possible to convert a problem with complicating variables to one with complicating constraints by disaggregating the y variables into the n variables y_i, $i = 1,\ldots,n$, and imposing the following constraints:

Figure 13.1 Problem with complicating variables.

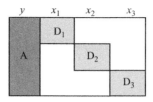

Figure 13.2 Problem with complicating constraints.

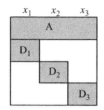

$$y_i = y_{i+1} \quad i = 1, 2, \ldots, n-1. \tag{13.3}$$

Hence, we concentrate only on Lagrangean decomposition because Benders decomposition, which can be viewed as an extension of the generalized Benders decomposition method that was presented in Chapter 9, will be covered in Chapter 14.

13.2 Lagrangean Relaxation

We start by introducing the concept of Lagrangean relaxation before we address Lagrangean decomposition. Major references on Lagrangean relaxation are Geoffrion (1974), Fisher (1981), Guignard and Kim (1987), Guignard (2003), and Frangioni (2005).

We will assume that we address the solution of MILP problems of the following form

$$
\begin{aligned}
(\mathbf{P}): \quad \min \quad & c^T x \\
\text{s.t.} \quad & Ax \leq b \\
& Ex \leq d \\
& x \in X \quad X = \{x_j = 0, 1 \ j = 1, \ldots, p \leq n, x_j \geq 0 \ j = p+1, \ldots, n\},
\end{aligned}
\tag{13.4}
$$

where the set X contains both 0-1 and continuous variables, and where $Ax \leq b$ are regarded as "complicating" constraints, while $Ex \leq d$ are "easy" constraints. The basic idea is that, if we remove the complicating constraints, the resulting MILP is relatively easy to solve. This may arise in MILPs with the decomposable structure shown in Fig. 13.2 when complicating constraints are removed, or in MILPs where the polyhedral structure of the "easy" constraints is favorable (e.g., the corresponding constraints have integer extreme points).

The basic idea in Lagrangean relaxation is to create an "easy" MILP by removing the "complicating" constraints. This can be accomplished by dualizing the objective function in which the complicating constraints are transferred into the objective function multiplied by the nonnegative Lagrange multipliers λ:

$$(\text{LR}_\lambda): \quad \min \quad c^T x + \lambda^T (Ax - b)$$
$$\text{s.t.} \quad Ex \leq d \qquad (13.5)$$
$$x \in X,$$

where $\lambda \geq 0$ are nonnegative Lagrange multipliers.

Property 13.1 The optimal value of the objective functions, $(\text{LR}_\lambda) \leq v(\text{P})$, which implies that (LR_λ) in (13.5) yields a lower bound to the original problem (P) in (13.4).

This follows from the fact that the feasible space is being relaxed by the removal of the complicating constraints $Ax \leq b$, and the fact that, as $\lambda^T (Ax - b) \leq 0$ for $\lambda \geq 0$, the objective function is being underestimated.

13.3 Lagrangean Dual

In order to find the tightest Lagrangean lower bound we consider the following problem, the Lagrangean dual, in which we determine the multipliers $\lambda \geq 0$ that maximize the lower bound

$$(\text{LR}): v(\text{LR}) = \max_{\lambda \geq 0} v(\text{LR}_\lambda). \qquad (13.6)$$

We should note that for MILP problems, the solution of the Lagrangean dual is such that $v(\text{LR}) \leq v(\text{P})$, meaning that no equality may be achieved, which is an indication that there might be a dual gap given the presence of integer variables in problem (P). Because there might be a dual gap, what is done in practice is to iterate on λ for a maximum number of iterations, which, strictly speaking, yields a heuristic solution if the gap is not reduced within a small tolerance ε.

We should note that if there are no integer variables in problem (P), it reduces to an LP, which we express in simplified form as the following primal problem (PP),

$$(\text{PP}): \quad \min \quad c^T x$$
$$\text{s.t.} \quad Ax \leq b \qquad (13.7)$$
$$x \geq 0.$$

If we consider the dual problem (DP) of the primal problem (PP), it is given by (Dantzig, 1963)

$$(\text{DP}): \quad \max \quad b^T \lambda$$
$$\text{s.t.} \quad A\lambda \geq c \qquad (13.8)$$
$$\lambda \geq 0.$$

The following property holds for the primal and dual problems.

Property 13.2 At the optimum solution of problems (PP) and (DP), $v(\text{PP}) = v(\text{DP})$.

What this means is that if there are no discrete variables, there is no dual gap because the optimum solution of the dual is then identical to the optimum solution of the primal problem.

For the MILP case, however, we can establish an interesting relation between the Lagrangean dual and the primal relaxation of problem (13.5). From MIP theory, Meyer's Theorem (Wolsey, 1998) establishes that there exists a set of inequalities,

$$Hx \leq h, \tag{13.9}$$

that, when added to the inequalities $Ex \leq d$, yields the convex hull of these inequalities,

$$\text{co}(x \in X, Ex \leq d) = (x \in X^R, Ex \leq d, Hx \leq h). \tag{13.10}$$

Let problem $(\tilde{P})$ be defined as follows:

$$
\begin{aligned}
(\tilde{P}) \quad \min \quad & c^T x \\
\text{s.t.} \quad & Ax \leq b \\
& \tilde{E}x \leq \tilde{d} \\
& x \in X^R,
\end{aligned}
\tag{13.11}
$$

where $Ex \leq d$, $Hx \leq h$ are replaced by $\tilde{E}x \leq \tilde{d}$, i.e., the convex hull of $Ex \leq d$ and X^R is the continuous relaxation of the integer and continuous variables specified in (13.4).

We can then establish the following theorem:

Theorem 13.1 $v(\text{LR}) = v(\tilde{P})$.

This is the optimal solution of the dual Lagrangean (LR) equal to the solution of problem $(\tilde{P})$. This is a very interesting result because it states that solving the dual Lagrangean is equivalent to having determined the convex hull of the noncomplicating constraints in problem (13.4). We present here only the outline of the proof (see Frangioni (2005) for rigorous proof).

Let $v(\tilde{P}) = \min_{x \in X^R} \left\{ c^T x \,|\, Ax \leq b, \tilde{E}^T x \leq \tilde{d} \right\}$. Consider its dual,

$$
\begin{aligned}
v(\tilde{P}) &= \max_{\lambda \geq 0, \mu \geq 0} \left\{ b^T \lambda + \tilde{d}^T \mu \,|\, A^T \lambda + \tilde{E}^T \mu \geq c \right\} \\
&= \max_{\lambda \geq 0} \left\{ b^T \lambda + \max_{\mu \geq 0} \left\{ \tilde{d}^T \mu : \tilde{E}^T \mu \geq c - A^T \lambda \right\} \right\} \\
&= \max_{\lambda \geq 0} \left\{ b^T \lambda + \min_{x \in X^R} (c^T - \lambda^T A)x : \tilde{E}x \leq \tilde{d} \right\} \\
&= \max_{\lambda \geq 0} \left\{ \min_{x \in X} \left\{ c^T x - \lambda^T (Ax - b) : Ex \leq d \right\} \right\} \\
&= v(\text{LR}).
\end{aligned}
\tag{13.12}
$$

Figure 13.3 Geometric illustration of relation of bounds of LP and dual of Lagrangean relaxation.

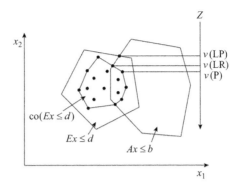

In this way we establish that the solution of the Lagrangean dual is equal to the primal relaxation with convex hull for $Ex \leq d$. Furthermore, we can establish the following corollary:

Corollary 1 $v(\text{LP}) \leq v(\text{LR})$.

This follows from the fact that the LP relaxation of (P), $v(\text{LP})$ is weaker than the relaxation of problem $(\tilde{\text{P}})$ with the convex hull for the noncomplicating variables, $v(\text{LP}) \leq v(\tilde{\text{P}})$. Hence, it follows that $v(\text{LP}) \leq v(\text{LR})$, i.e., the dual of the Lagrangean yields lower bounds greater or equal to the LP relaxation (see Fig. 13.3). Note also that $v(\text{LP}) = v(\text{LR})$ if $Ex \leq d$ defines the convex hull of these constraints.

As an example of the application of Lagrangean relaxation consider the generalized assignment problem for machines $i = 1, \ldots, m$ and jobs $j = 1, \ldots, n$, where c_{ij} are the corresponding cost coefficients, and the coefficients a_{ij} (0 or 1) and b_i determine the feasibility of assignment:

$$
\begin{aligned}
\min \quad & \sum_i \sum_j c_{ij} x_{ij} \\
\text{s.t.} \quad & \sum_j a_{ij} x_{ij} \leq b_i \quad \forall i \\
& \sum x_{ij} = 1 \qquad \forall j \\
& x_{ij} = 0, 1.
\end{aligned}
\tag{13.13}
$$

The first set of constraints can be interpreted as knapsack constraints, while the second set of constraints are multiple-choice constraints (each job j can only be assigned to one machine i). Say that we dualize these multiple-choice constraints as follows:

$$
\begin{aligned}
\min \quad & \sum_i \sum_j c_{ij} x_{ij} + \sum_j \lambda_j \left(1 - \sum_i x_{ij} \right) \\
\text{s.t.} \quad & \sum_j a_{ij} x_{ij} \leq b_i \qquad \forall i \\
& x_{ij} = 0, 1.
\end{aligned}
\tag{13.14}
$$

Model (13.14) can be rearranged as follows, where (13.15a) and (13.15b) are equivalent:

$$\min \quad \sum_i \sum_j \left(c_{ij} - \lambda_j\right) x_{ij} + \sum_j \lambda_j$$
$$\text{s.t.} \quad \sum_j a_{ij} x_{ij} \leq b_i \qquad \forall i \tag{13.15a}$$

$$\sum_i \min \quad \left(\sum_j c_{ij} - \lambda_j\right) x_{ij} + \sum_j \lambda_j$$
$$\text{s.t.} \quad \sum_j a_{ij} x_{ij} \leq b_i \qquad \forall i. \tag{13.15b}$$

Problem (13.15b) can then be solved independently for each machine i. Furthermore, this means that for fixed λ_j, we have for each machine i a knapsack problem,

$$Z^i = \min \quad \sum_j \left(c_{ij} - \lambda_j\right) x_{ij}$$
$$\text{s.t.} \quad \sum_j a_{ij} x_{ij} \leq b_i \tag{13.16}$$
$$x_{ij} = 0, 1.$$

The lower bound is then given by

$$v(\mathbf{LR}_\lambda) = \sum_i Z^i + \sum_j \lambda_j \leq v(\mathbf{P}). \tag{13.17}$$

Because the knapsack constraints do not yield the convex hull of the feasible region, this then implies that the lower bound from the Lagrangean dual is stronger than the lower bound of the LP relaxation, that is $v(\mathbf{LR}) > v(\mathbf{LP})$. Before we address the issue of how to readjust the λs to maximize $v(\mathbf{LD})$, we will consider Lagrangean decomposition.

13.4 Lagrangean Decomposition

Assume we reformulate problem (P) in (13.4) by copying the x variables into the y variables (Guinard and Kim, 1987) so that we assign these variables to the two different types of constraints, complicating and noncomplicating, that is,

$$(\mathbf{P'}) : \quad \min \quad c^T x$$
$$\text{s.t.} \quad Ax \leq a$$
$$Ey \leq d \tag{13.18}$$
$$x = y$$
$$x \in X, \quad y \in Y.$$

We should note that this reformulation is especially useful for problems with staircase matrices. In a similar way as we did in the previous section, we can define a dual Lagrangean decomposition $(\mathbf{LD}_\lambda)$:

$$
\begin{aligned}
(\mathbf{LD}_\lambda): \quad \min \quad & c^T x + \lambda^T (x - y) \\
\text{s.t.} \quad & Ax \leq a \\
& Ey \leq d \\
& x \in x, \quad y \in Y.
\end{aligned}
\tag{13.19}
$$

For fixed value of λ (in this case they are unconstrained in sign due to the equality $x = y$), (13.19) decomposes into two subproblems: one in the space of x and the other in the space of y,

$$
\begin{aligned}
\min \quad & \frac{1}{2} c^T x + \lambda^T x \\
\text{s.t.} \quad & Ax \leq a \\
& x \in X
\end{aligned}
\tag{13.20}
$$

$$
\begin{aligned}
\min \quad & \frac{1}{2} c^T y - \lambda^T y \\
\text{s.t.} \quad & Ey \leq d \\
& y \in Y,
\end{aligned}
\tag{13.21}
$$

where we have arbitrarily split the linear term $c^T y$ in equal parts between the two subproblems. Similar to Theorem 13.1, we can prove that

$$
\begin{aligned}
(\tilde{P}') = \min \quad & c^T x \\
\text{s.t.} \quad & x \in \mathrm{co}(x \in X, Ax \leq b) \cap \mathrm{co}(x \in X, Ex \leq d),
\end{aligned}
\tag{13.22}
$$

which implies $v(\tilde{P}') = v(\mathbf{LD}) \geq v(\mathbf{LR})$. Thus, the lower bound of Lagrangean decomposition yields stronger lower bounds than the lower bound from Lagrangean relaxation (see Fig. 13.4).

Figure 13.4 Geometric illustration of relation of bounds of LP and dual of Lagrangean decomposition.

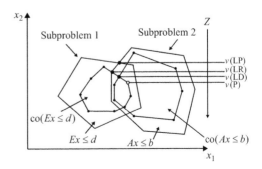

13.5 Update of Lagrange Multipliers

In order to update the Lagrangean relaxation or Lagrangean decomposition, we would ideally like to select the multipliers to maximize the corresponding dual functions. For simplicity in the presentation, we consider the updates of the multipliers λ in the dual of the Lagrangean relaxation LR_λ

$$\begin{aligned} \min \quad & c^T x + \lambda^T (Ax - b) \\ \text{s.t.} \quad & Ex \le e \\ & x \in X. \end{aligned} \tag{13.23}$$

Assume that the convex hull of the inequalities $Ex \le e$, is a bounded polyhedron with extreme points x^k, $k \in V$. We can then rewrite LR_λ as follows:

$$\min_{k \in V} \left\{ c^T x^k + \lambda^T \left(Ax^k - b \right) \right\} = Z(\lambda). \tag{13.24}$$

The geometric interpretation of problem (13.24) is shown in Fig. 13.5, where it can be seen that $Z(\lambda)$ is a concave function of λ.

Since we want to determine $\max_\lambda v(LR_\lambda)$, we can rewrite $Z(\lambda)$ as follows:

$$\begin{aligned} Z(\lambda) = \ & \max \eta \\ \text{s.t.} \quad & \eta \le c^T x^k + \lambda^T \left(Ax^k - b \right) \quad \forall k \in V. \end{aligned} \tag{13.25}$$

The drawback with the LP (13.25) is that it requires knowing all the extreme points, x^k, $k \in V$, and hence it is impractical to set up such a problem. To circumvent this problem, consider a relaxation in which we only consider a subset of known extreme points, $k \in \bar{V} \subset V$. This LP is the basis of the cutting-plane method for updating the Lagrange multipliers (Guignard, 2003).

Because the cutting-plane method is initially slow until it accumulates enough cutting planes, an alternative method is the subgradient method (Fisher, 1981). In general, we have

Figure 13.5 Piecewise linear approximation of Lagrangean dual.

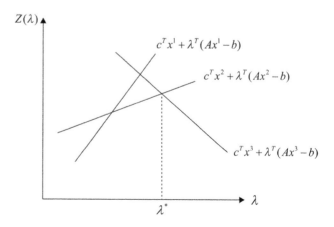

that a subgradient γ is given by the following inequality, in which the right-hand side overestimates the change in function value from x to x^0,

$$f(x) - f(x^0) \leq \gamma(x - x^0). \tag{13.26}$$

From (13.24) we can define the subgradient as

$$s^k = Ax^k - b. \tag{13.27}$$

Let the new multiplier be given by this subgradient multiplied by the scalar μ, which can be interpreted as a step size in analogy to the steepest-descent method,

$$\lambda^{k+1} = \lambda^k + \mu s^k. \tag{13.28}$$

Because we have

$$c^T x^k + (\lambda^k)^T (Ax^k - b) = \eta^k, \tag{13.29a}$$

and if we assume

$$c^T x^k + (\lambda^{k+1})^T (Ax^k - b) = \eta^*, \tag{13.29b}$$

from the definition of subgradient in (13.27) and from (13.29a) and (13.29b), we can set,

$$\eta^* - \eta^k = (s^k)^T (\lambda^{k+1} - \lambda^k). \tag{13.30}$$

Substituting (13.30) in (13.28), yields

$$\eta^* - \eta^k = \mu (s^k)^T s^k, \tag{13.31}$$

or

$$\mu = \frac{\eta^* - \eta^k}{(s^k)^T s^k}. \tag{13.32}$$

From (13.28), and because η^* is unknown, we replace it by the upper bound Z^U

$$\lambda^{k+1} = \lambda^k + \frac{\alpha_k (Z^U - \eta^k)}{\|s^k\|^2} s^k, \tag{13.33}$$

where $\eta^k = c^T x^k + (\lambda^k)^T (Ax^k - b)$ and the step size α_k is typically chosen in the interval [0,2], that is $\alpha_k \in [0, 2]$.

In summary, the algorithm for Lagrangean relaxation consists of the following steps.

(1) Guess initial value of the multipliers λ^k (e.g., from the relaxed LP).
(2) Solve (LR$_\lambda$) to yield lower bound $v(\text{LR}_\lambda)$.
(3) Compute from (13.18) the upper bound Z^U using rounding heuristics.
(4) If $|v(\text{LR}_\lambda) - Z^U| < \varepsilon$, stop.

Otherwise update λ^{k+1} using (13.25) or (13.33) and go to step (2).

Notice that if we were to apply the above steps to the Lagrangean decomposition, then step (2) decomposes into subproblems by solving the dual Lagrangaen LD_λ.

Also, we can make the following remarks.

(1) Owing to the dual gap, the specified tolerance ε may not be satisfied. Therefore, we specify a maximum number of iterations as a termination criterion. We can also use a branch and bound method to close the gap (e.g., see Carøe and Schultz, 1999).

(2) We can also embed the LR_λ or LD_λ in a rigorous branch and bound search to compute lower bounds that are stronger than the LP relaxation.

(3) If there are equalities in problem (13.4), the multipliers λ in the Lagrangean decompositions are unrestricted in sign.

(4) The Lagrangean relaxation or decomposition can be extended to the solution of NLP and MINLP problems.

EXERCISES

13.1 Discuss the relative advantages and disadvantages of applying Lagrangean relaxation versus Lagrangean decomposition to the solution of the MILP problem (13.4). Are there any special cases for which both decompositions will yield the same computational performance?

13.2 Consider the following multiperiod LP problem:

$$\min \quad Z = c^T y + \sum_{i=1}^{n} a_i^T x_i$$
$$\text{s.t.} \quad Cy + Ax_i = b, \qquad i = 1, 2, \ldots, n$$
$$y \geq 0, x_i \geq 0, \qquad i = 1, 2, \ldots, n,$$

where n is the number of time periods.

(a) Reformulate the problem so that the LP can be solved by Lagrangean decomposition.

(b) Show the explicit form of the Lagrangean decomposition problem.

14 Stochastic Programming

14.1 Strategies for Optimization under Uncertainty

In optimization models there is often the need to anticipate effects of uncertainty in the data (e.g., demands in planning and scheduling applications) (Sahinidis, 2004). The problem of optimization under uncertainty (assuming only inequalities) is generally given in initial form as follows:

$$\min \quad f(x, \theta)$$
$$\text{s.t.} \quad g(x, \theta) \leq 0 \tag{14.1}$$
$$x \in X,$$

where x are the decision variables and θ is the vector of uncertain parameters (e.g., demands). Problem (14.1) is not a well-defined problem, as we need to specify a strategy for hedging against the uncertainty. One approach is to adopt a deterministic strategy where, for instance, we assume that the uncertain parameters are specified in a given uncertainty set, $\theta \in T$, e.g., $T = \{\theta^L \leq \theta \leq \theta^u\}$. The idea is then to specify a given criterion on how to anticipate the effect of the parameters θ. One possible criterion is to adopt a minimax strategy in which we optimize the objective function for the worst outcome of θ, that is,

$$\min_{x \in X} \max_{\theta \in T} \{f(x, \theta) | g(x, \theta) \leq 0\}. \tag{14.2}$$

The solution to problem (14.2) can then be interpreted as yielding the decision x that is optimal for the "worst" condition for $\theta \in T$.

An alternative criterion is to use a robust optimization strategy in which the concern is to select the variable x so as to ensure the feasibility of the inequalities $g(x, \theta) \leq 0$ for all θ in T. Mathematically that leads to the problem

$$\min \quad f(x)$$
$$\text{s.t.} \quad g(x, \theta) \leq 0 \quad \forall \theta \in T \tag{14.3}$$
$$x \in X,$$

which is known as a semi-infinite programming problem (Hettich and Kortanek, 1993) since an infinite number of inequalities $g(x, \theta) \leq 0$ have to be satisfied for all points θ in T. Problem (14.3) can be reformulated as a finite-dimensional optimization problem by maximizing each individual constraint $g_i(x, \theta)$, $i \in I$, with respect to θ. That is,

$$\min \ f(x)$$

$$\text{s.t.} \quad \max_{\theta \in T} \ g_i(x, \theta) \leq 0 \quad \forall i \in I \tag{14.4}$$

$$x \in X.$$

Finally, if we are given a probability distribution function (pdf) $\phi(\theta)$ we can also formulate the chance-constrained optimization problem (Prékopa, 1970)

$$\min \ f(x)$$

$$\text{s.t.} \quad \text{prob} \ (g(x, \theta) \leq 0) \geq 1 - \alpha \tag{14.5}$$

$$x \in X$$

in which we require the inequalities to be satisfied with a minimum probability $1 - \alpha$. Notice that the operator *prob* in (14.5) implies the integration of a multiple integral in θ over the feasible region $g(x, \theta) \leq 0$, a nontrivial problem. For this reason, problem (14.5) is often replaced by individually imposing probability conditions on each constraint,

$$\min \ f(x)$$

$$\text{s.t.} \quad \text{prob} \ (g_i(x, \theta) \leq 0) \geq 1 - \alpha_i, i \in I \tag{14.6}$$

$$x \in X,$$

in which the probabilistic constraints are reformulated by taking the inverse of the cumulative distribution function, $\hat{F}$,

$$\min \ f(x)$$

$$\text{s.t.} \ g_i(x, \theta) \leq \hat{F}^{-1}(1 - \alpha_i), i \in I, \tag{14.7}$$

which leads to a much more manageable problem.

An alternative approach to account for uncertainty in optimization is stochastic programming (Birge and Louveaux, 2011). Here we also assume that the uncertain parameters θ are described by a pdf, $\phi(\theta)$, which can be continuous or discrete. There are different ways of defining the objective function and treatment of the variables, as follows.

(a) Minimization of expected cost

$$\min_{x \in X} \int .. \int [f(x, \theta) | g(x, \theta) \leq 0] \phi(\theta) d\theta \tag{14.8}$$

in which we select the variables x so as to minimize the average or expected cost, i.e.,

$$\min_{x \in X} \ \underset{\theta}{E} \ [f(x, \theta) | g(x, \theta) \leq 0]. \tag{14.9}$$

If the pdf is continuous in θ, this involves selecting x to minimize the multiple integral in (14.8). In contrast, if we have a discrete distribution function where the discrete uncertainties θ_i, $i = 1, \ldots, n$, each have a probability given by $p_i = \phi(\theta_i)$, the objective involves a summation of the objective weighted by the probabilities,

$$\min_{x \in X} \sum_{i=1}^{n} p_i f(x, \theta_i)$$

$$\text{s.t.} \quad g(x, \theta_i) \leq 0, i = 1, \ldots, n$$

$$x \in X.$$

(14.10)

(b) Two-stage programming

The basic idea here is to recognize that in many problems the decisions x can be split into two: stage-1 decisions d that are "here and now" in the sense that once selected they cannot be changed or adjusted, and stage-2 decisions z that are "wait and see" in that these decisions can be adjusted according to the realization of the uncertain parameters θ. These variables are also often denoted as design decisions for stage 1, and recourse decisions for stage 2. The basic idea is illustrated in Fig. 14.1. As an example, consider a supply-chain problem. We can regard the selection of the structure and process capacities as stage-1 decisions, while stage-2 decisions could be levels of production that are adjusted according to the uncertain demands.

If we partition the vector of variables x into design variables d (stage 1) and control or recourse variables z,

$$x = \begin{bmatrix} d \\ z \end{bmatrix},$$

we will assume that the recourse variable z can be adjusted for each realization of the uncertain parameters θ, as shown in Fig. 14.2, which implies that one can measure the realization of θ and make immediate adjustment of z (e.g., perfect control).

To formulate the two-stage programming problem, we consider first the stage-2 subproblem where for fixed stage-1 decision d, we optimize z for each parameter value θ:

$$\min_{z} \quad f(d, z, \theta)$$

$$\text{s.t.} \quad g(d, z, \theta) \leq 0,$$

(14.11)

Figure 14.1 Illustration for two-stage stochastic programming.

d (design)   z (recourse)

Stage 1: here and now   Stage 2: wait and see

Figure 14.2 For fixed stage-1 d, recourse z can be adjusted for each θ.

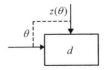

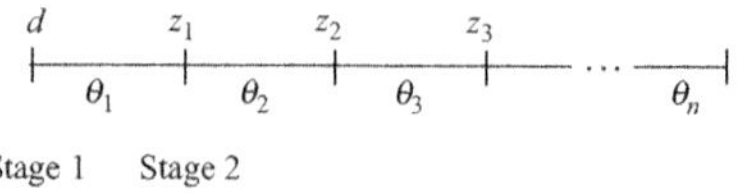

Figure 14.3 Multistage stochastic programming.

if we take the average for all θ, the expected value, $E_\theta\left[\min_z f(d, z, \theta)|g(d, z, \theta) \le 0\right]$, then selecting the stage-1 variable d and stage-2 variable $z(\theta)$ leads to the two-stage programming problem, which can be formulated as:

$$\min_d E_\theta \left[\min_z f(d, z, \theta)\Big|g(d, z, \theta) \le 0 \right]. \tag{14.12}$$

We should note that the formulation in (14.12) implicitly assumes feasibility $\forall \theta$ for the two-stage problem in (14.11), which is known as full recourse. To allow for the possibility of violation of constraints, a slack variable u_i with a corresponding penalty cost c_i can be introduced for each inequality, with which the two-stage programming problem can be formulated as follows:

$$\min_d \left[E_\theta \left[\min_z f(d, z, \theta) + \sum c_i u_i \Big| g_i(d, z, \theta) \le u_i, u_i \ge 0 \forall i \right] \right]. \tag{14.13}$$

(c) Multistage stochastic programming

The two-stage programming problem can be extended to multiple stages of decision as shown in Fig. 14.3, in which the recourse variables z_k in a given stage k, become stage-k variables for stage $k + 1$. The multistage stochastic programming problem can be formulated with nested expectations as follows:

$$\min_d \left[E_{\theta_1} \min_{z_1} \left[E_{\theta_2} \min_{z_2} \left[E_{\theta_3} \min_{z_3} f(d, z_1, z_2, z_3, \theta_1, \theta_2, \theta_3) \right. \right. \right.$$
$$\left. \left. \left. \text{s.t.} \quad g[d, z_1, z_2, z_3, \theta_1, \theta_2, \theta_3) \le 0 \right] \right] \right]. \tag{14.14}$$

14.2 Linear Stochastic Programming

Consider the two-stage LP stochastic programming problem

$$\begin{aligned} \min \quad & c^T x + \bar{Q}(x) \\ \text{s.t.} \quad & Ax = b \\ & x \ge 0, \end{aligned} \tag{14.15}$$

where x are the stage-1 variables, and $\bar{Q}(x)$ is the expected recourse function

$$\bar{Q}(x) = E_\theta\{Q(x, \theta)\}, \tag{14.16}$$

Figure 14.4 Expected recourse function $\bar{Q}(x)$.

where for every realization of the vector of uncertain parameter θ that obeys a given probability distribution function, $Q(x, \theta)$ is optimized in terms of the stage-2 variables y, which are expressed as $y(\theta)$ to denote the fact that these variables are adjusted for each value of θ

$$Q(x, \theta) = \min q(\theta)^T y(\theta)$$
$$\text{s.t.} \quad W(\theta)y(\theta) = h(\theta) - T(\theta)x \tag{14.17}$$
$$y(\theta) \geq 0.$$

In (14.17) the uncertainties can arise in the cost coefficient $q(\theta)$, in the right-hand side $h(\theta)$, in the matrix $W(\theta)$ of the recourse (stage-2) variables, and in the matrix $T(\theta)$ (sometimes known as the "Technology Matrix") of the stage-1 variables.

The following property can be established for the expected recourse function.

Property 14.1 The function $\bar{Q}(x)$ is convex and piecewise linear.

The above property follows from the fact that the subproblem in (14.15) is linear, and the fact that changes in the basis of the LP give rise to breakpoints in the piecewise linear function that is shown in Fig. 14.4.

From (14.16) and (14.17) the two-stage stochastic programming in (14.15) can be written as follows:

$$\min \quad c^T x + E_\theta \left\{ \min_y \ q(\theta)^T y(\theta) \right\}$$

$$\text{s.t.} \quad Ax = b \tag{14.18}$$
$$W(\theta)y(\theta) = h(\theta) - T(\theta)x$$
$$x \geq 0 \quad y(\theta) \geq 0.$$

If the uncertain parameters θ are described by a discrete probability distribution function, θ_j, $j = 1, \ldots, n$, with probabilities $p_j, j = 1, \ldots, n$ (see Fig. 14.5), (14.18) can be expressed as the finite-dimensional linear programming problem

$$\min \quad c^T x + \sum_j p_j q_j^T y_j$$

$$\text{s.t.} \quad Ax = b \tag{14.19}$$
$$T_j x + W_j y_j = h_j \quad j = 1, \ldots, n$$
$$x \geq 0 \quad y_j \geq 0,$$

Figure 14.5 Two-dimensional discretization of two uncertain parameters.

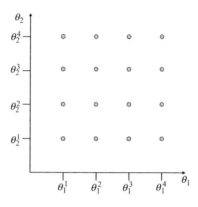

Figure 14.6 Scenario tree for three parameter realizations.

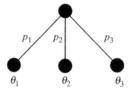

where q_j, h_j, W_j, and T_j are evaluated at the discrete points θ_j, $j = 1, \ldots, n$. Each of these points can be interpreted as a scenario that is realized in stage 2, as shown in the scenario tree in Fig. 14.6.

The LP in (14.19) is known as the deterministic equivalent problem of the two-stage programming problem (14.18), and can simply be solved as an LP. If some of the stage-1 and stage-2 variables are integers, the deterministic equivalent in (14.19) gives rise to an MILP problem. In either case, the difficulty might be that the size of the deterministic equivalent is too large if the number of scenarios is large. To address this problem either a sampling or a decomposition method is required (Birge and Louveaux, 2011). We present below the Benders decomposition method to address this problem.

14.3 L-Shaped Method

Benders decomposition (Benders, 1962) as applied to the two-stage LP problem in (14.18) is also known as the L-shaped method (Van Slyke and Wets, 1969). The stage-1 variables x can be regarded as complicating variables, so that if they are fixed at the value x^k at iteration k, the problem decomposes in the scenarios $j = 1, \ldots, n$ that can be solved independently in parallel (see Fig. 14.7). Specifically, the L-shaped method consists of solving a master problem in terms of the stage-1 variables x, and the subproblems $j = 1, \ldots, n$ for the stage-2 variables y_j for fixed variables x, as shown in Fig. 14.8.

Figure 14.7 Block diagonal structure for two-stage programming.

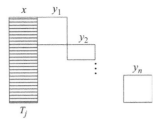

Figure 14.8 The L-shaped method (Benders decomposition).

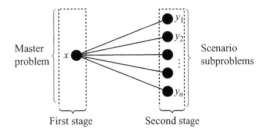

From (14.19) it follows that the subproblem $\left(S_j^k\right)$ of scenario j at iteration k is given by:

$$Z_j^k = \min \ q_j^T y_j$$
$$\text{s.t.} \ \ W_j y_j = h_j - T_j x^k \tag{14.20}$$
$$y_j \geq 0.$$

Consider the Lagrangean function of the j subproblem (14.20) at iteration k,

$$L_j^k = q_j^T y_j + \left(\lambda_j^k\right)^T \left[W_j y_j - h_j + T_j x^k\right], \tag{14.21}$$

where λ_j^k are the KKT multipliers of the equality constraints. The stationary conditions of (14.21) yield

$$\frac{\partial L_j^k}{\partial y_j} = q_j + W_j^T \lambda_j^k = 0. \tag{14.22}$$

Given a total of K iterations, we can define the master problem for problem (14.19) as follows:

$$\min \ \eta$$
$$\text{s.t.} \ \ \eta \geq c^T x + \sum_j p_j L_j^k \left(x, y_j, \lambda_j^k\right), k = 1, \ldots, K$$
$$Ax = b \tag{14.23}$$
$$x \geq 0, \eta \in R^1.$$

Substituting (14.21) into (14.23) yields

$$\min \ \eta$$
$$\text{s.t.} \quad \eta \geq c^T x + \sum_j p_j \left[q_j^T y_j + \sum_j \left(\lambda_j^k \right)^T \left[W_j y_j + T_j x - h_j \right] \right] \quad k = 1, \ldots, K \tag{14.24}$$
$$Ax = b$$
$$x \geq 0, \eta \in R^1,$$

which can be simplified with (14.22) as follows,

$$\min \ \eta$$
$$\text{s.t.} \quad \eta \geq c^T x + \sum_j p_j \left(\lambda_j^k \right)^T T_j x - \sum_j p_j \left(\lambda_j^k \right)^T h_j \quad k = 1, \ldots, K \tag{14.25}$$
$$Ax = b$$
$$x \geq 0, \eta \in R^1.$$

Note that the master problem in (14.25) is expressed in terms of the stage-1 variables x and the objective η, and therefore it can be regarded as a projection that corresponds to an approximation of the piecewise linear function in Fig. 14.4. Furthermore, its solution yields a lower bound to the two-stage LP problem (14.19). Each inequality in (14.25) is denoted as an optimality cut, where it is assumed that the subproblems $\left(S_j^k \right)$ are feasible. In case any subproblem is infeasible one can derive a feasibility cut that results from minimizing the violation in (14.20) (using 1-norm as discussed below). The feasibility cut for a given iteration k is given by the inequality

$$\sum_j \left(\mu_j^k \right)^T \left[T_j x - h_j \right] \leq 0, \tag{14.26}$$

where μ_j^k are the multipliers of the feasibility problem.

In summary, by defining the set FEA as those iterations that yield feasible subproblems, and the set INF as those iterations with infeasible subproblems, the master problem M^k at iteration k is given by the following LP,

$$\text{LB}^k = \min \ \eta$$
$$\text{s.t.} \quad \eta \geq c^T x + \sum_j p_j \left(\lambda_j^k \right)^T T_j x - \sum_j p_j \left(\lambda_j^k \right)^T h_j, \quad k \in \text{FEA}$$
$$\sum_j \left(\mu_j^k \right)^T \left[T_j x - h_j \right] \leq 0 \quad k \in \text{INF} \tag{14.27}$$
$$Ax = b$$
$$x \geq 0, \quad \eta \in R^1,$$

where LB^k corresponds to the lower bound predicted at each iteration k. The upper bound UB^k is then given by

$$UB^k = \sum_j p_j Z_j^k + c^T x^k, \tag{14.28}$$

where Z_j^k is the solution to the subproblem in (14.20).

The L-shaped or Benders decomposition algorithm is then given by the following steps.

(1) Set $k = 0$.
(2) Solve the master problem M^k in (14.27) to obtain x^k and the lower bound LB^k.
(3) For a fixed value of the stage-1 variable x^k, solve the subproblems $\left(S_j^k\right) j = 1, \ldots, n$ in (14.20) to obtain the multipliers λ_j^k for the optimality cuts. If the subproblems are feasible, the upper bound is given by $UB^k = \sum_j p_j Z_j^k + c^T x^k$. If any of the subproblems is infeasible, solve the feasibility problem,

$$\min \quad v^+ - v^-$$
$$\text{s.t.} \quad W_j y + I v^+ - I v^- = h_j - T_j x^k$$
$$y, v^+, v^- \geq 0,$$

to obtain the multipliers μ_j^k. Set the upper bound to $UB^k = \infty$.
(4) Iterate until $LB^k \simeq \min_k \left\{ UB^k \right\}$.

Figure 14.9 shows typical iterations with Benders decomposition, where, as it can be seen, the lower bounds, LB, increase monotonically because cumulative Benders cuts are added to the master problem at each iteration. In contrast, the upper bounds, UB, do not follow a decreasing pattern as trial points for the UB are generated at each iteration, which may lead to improvements, or not, of the upper bound.

We should note that the above algorithm can be extended to the multicuts version, where cuts in the master problem are added for each scenario subproblem at each iteration (see Exercise 14.1). It can be shown that this yields stronger lower bounds, but at the expense of solving a large master problem (Birge and Louveaux, 2011).

We should also note that the above algorithm can be applied when the stage-1 variable x has 0-1 variables, in which case the master problem gives rise to an MILP while the subproblems are still LPs. The extension to stage-2 variable y with 0-1 variables is not trivial, as then one cannot apply the dual conditions in (14.21), which then gives rise to a dual gap (Schultz, 2003). Finally, for a computationally efficient implementation, the solution of the subproblems $\left(S_j^k\right) j = 1, \ldots, n$ in (14.19) can be parallelized because they are independent of each other.

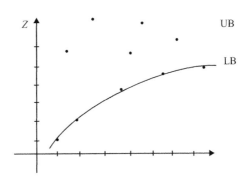

Figure 14.9 Progress of iterations in Benders decomposition.

14.4 Multistage Stochastic Programming

As indicated at the end of Section 14.1, two-stage programming problems can be extended to multiple stages of decision as shown in Fig. 14.3, in which the recourse variables z_k, in a given stage k, become stage-k variables for stage $k + 1$. Figure 14.10 shows the scenario tree of a three-stage programming problem in which there are two realizations of the uncertain parameters at each stage. As can be seen in Fig. 14.10, there are four terminal nodes, meaning a total of four scenarios.

The mathematical formulation for the three-stage programming problem in Fig. 14.10 is not trivial to develop, because one has to write the model in terms of parent and children nodes in the scenario tree. However, one can replace the tree representation in Fig. 14.10 by the alternative representation (Ruszczynski, 1997) shown in Fig. 14.11, in which each of the scenarios Sc1, Sc2, Sc3, and Sc4 are represented by independent paths each with variables, x, y, and z, associated to each scenario. In order to map this representation into the original tree in Fig. 14.10, horizontal links between nodes are imposed to equate the variables involved in the interconnected nodes. So, for instance, at stage 1, the links between the four nodes of the scenarios represent the equality of the variables at those nodes, namely, $x^1 = x^2 = x^3 = x^4$, which correspond to the root node of the tree in Fig. 14.10. Similarly, at stage 2, the link between the nodes of scenarios Sc1 and Sc2, represents the equality $y^1 = y^2$, which represents the first node at stage 2 in Fig. 14.10. The equality constraints that are associated to the links in Fig. 14.11 are known as nonanticipativity constraints (Ruszczynski, 1997).

Based then on the nonanticipativity constraints, $x^1 = x^2$, $x^2 = x^3$, $x^3 = x^4$, $y^1 = y^2$, and $y^3 = y^4$, the multistage stochastic programming problem can be written as follows:

Figure 14.10 Scenario tree for a three-stage problem.

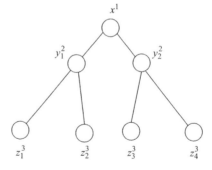

Figure 14.11 Alternative representation for multistage tree.

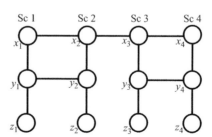

$$\min \sum_s p_s \left(c^T x^s + q_1^T y + q_2^T z^s \right)$$

$$
\begin{aligned}
\text{s.t.} \quad & Ax^s = b, && \forall s \\
& T_1 x^s + W_1 y^s = h_1^s, && \forall s \\
& T_2 y^s + W_2 z^s = h_2^s, && \forall s \\
& x^1 = x^2 = x^3 = x^4 \\
& y^1 = y^2, y^3 = y^4 \\
& x^s, y^s, z^s \geq 0 \; \forall s.
\end{aligned}
\tag{14.29}
$$

Formulation (14.29) represents the deterministic equivalent to the three-stage stochastic programming problem. Since this problem can become very large, it can be solved with Lagrangean decomposition (see Chapter 13) by dualizing the nonanticipativity constraints, as then problem (14.29) can be solved independently for each scenario (Ruszczynski, 1997).

14.5 Robust Optimization

We should also note that, in terms of applications of optimization under uncertainty, the alternative approach to stochastic programming is robust optimization (El Ghaoui, 1997; Bertsimas et al., 2011) in which the major concern is to ensure feasibility over a specified uncertainty set as formulated in (14.3).

For the specific case of linear programs, which is the most common model for robust optimization, we consider the following model,

$$
\begin{aligned}
\min \quad & \sum_j \tilde{c}_j x_j \\
\text{s.t.} \quad & \sum_j \tilde{a}_{ij} x_j \leq \tilde{b}_i \; \forall i \in I \\
& x \geq 0,
\end{aligned}
\tag{14.30}
$$

which involves uncertainty in the cost coefficients $\tilde{c}_j$, left-hand side coefficients $\tilde{a}_{ij}$, and right-hand sides $\tilde{b}_i$. For convenience we rearrange (14.30) as follows:

$$
\begin{aligned}
\min \quad & z \\
\text{s.t.} \quad & z - \sum_j \tilde{c}_j x_j \leq 0 \\
& \tilde{b}_i x_0 + \sum_j \tilde{a}_{ij} x_j \leq 0 \; \forall i \in I \\
& x \geq 0, x_0 = -1,
\end{aligned}
\tag{14.31}
$$

which in turn can be reduced to the compact form

$$\min \quad f^T x$$
$$\text{s.t.} \quad A(\xi)x \le b \ \forall \xi \in \Xi \tag{14.32}$$
$$x \le 0,$$

where the inequalities $A(\xi)x \le b$ correspond to the two inequalities in (14.31), ξ are the uncertain parameters, and the linear objective function generalizes the term z.

If we consider a polyhedral set for the P uncertain parameters,

$$\Xi = \left\{ \xi : V\xi \le u, \xi \in R_+^P \right\}. \tag{14.33}$$

we can express (14.32) as the semi-infinite programming problem:

$$\min \quad f^T x$$
$$\text{s.t.} \quad (a_i + P_i\xi)^T x \le b_i \ \forall i \in I, \xi \in \Xi \tag{14.34}$$
$$x \ge 0.$$

As will be shown in Chapter 15, the feasibility of the constraints in (14.33) can be established by maximizing each constraint i,

$$a_i^T + \max_{\xi \in \Xi} (P_i\xi)^T x \le b_i. \tag{14.35}$$

Taking the dual of the inner maximization problem,

$$\max_{\xi \in \Xi} \left(P_i^T x \right)^T \xi$$
$$\text{s.t.} \quad V\xi \le u, \tag{14.36}$$

yields

$$\min_{\lambda \in R_+^m} u^T \lambda_i$$
$$\text{s.t.} \quad V^T \lambda_i \ge P_i^T x. \tag{14.37}$$

Substituting (14.36) in (14.34) and next in (14.33), and by omitting the min operation (assuming weak duality), leads to the robust counterpart formulation that yields the LP (Ben-Tal and Nemirovsky, 1999),

$$\min \quad f^T x$$
$$\text{s.t.} \quad a_i^T x + u_i^T \lambda_i \le b_i \ \forall i \in I$$
$$\quad V^T \lambda_i \ge P_j^T x \qquad \forall i \in I \tag{14.38}$$
$$\quad \lambda_i \in R_+^m \qquad\qquad \forall i \in I, \ x \ge 0.$$

Problem (14.38) can be solved efficiently as it is an LP. We should also note that it is possible to consider other uncertainty sets (e.g., ellipsoidal, cone), and account for restricted classes of recourse variables (e.g., adjustable linear functions) (Ben-Tal et al., 2004).

Finally, as to whether stochastic programming or the robust approach is more suitable depends on the problem at hand. For instance, in the area of process operations, stochastic programming (two stages or multistage) is generally more meaningful for planning problems due to their longer time horizons in which recourse decisions can be fully anticipated (Alonso-Asuyo et al., 2003). On the other hand, for scheduling problems, robust optimization is generally more meaningful since the major concern tends to be to satisfy production demands over short-term horizons in which recourse actions may not be readily applied (Lin et al., 2004).

EXERCISES

14.1 Rewrite the steps of the L-shaped algorithm for the case when optimality or feasibility cuts for each scenario j are added to the master problem.

14.2 Consider the mixed-integer linear programming problem

$$\min \quad Z = a^T x + b^T y$$
$$\text{s.t.} \quad Ax + By \leq d$$
$$x \geq 0, y \in \{0,1\}.$$

Assume it is desired to solve this problem by Benders decomposition where the 0-1 variables are treated as "complicating" variables for the master problem. If the **LP** subproblems for fixed y^k, $k = 1, 2, \ldots, K$ are feasible with an optimal solution x^k and multipliers λ^k, show that the master problem can be formulated as follows:

$$Z_L^K = \min \alpha$$
$$\text{s.t. } \alpha \geq b^T y + \lambda_k^T [By - d], \quad k = 1, 2, \ldots, K$$
$$\alpha \in R^1, \quad y \in \{0,1\}.$$

15 Flexibility Analysis

15.1 Introduction

In this chapter, we address one of the important components in the operability of a process system, namely, flexibility (see Grossmann et al., 2014, for a general review). By flexibility we mean the capability of a system design to have feasible steady-state operation for a range of uncertain conditions that may be encountered during plant operation. Clearly, there are other aspects to the operability of a plant, such as controllability, safety, and reliability that are equally important. However, flexibility is a general attribute as it provides a characterization for the capability of a design to ensure feasibility in the face of uncertainty while anticipating that corrective actions can be taken to ensure feasibility. For an interesting relationship between flexibility and robust optimization see Zhang et al. (2016).

In this chapter we will concentrate on two basic analysis problems for flexibility. The first one will be the problem when we need to determine if a design is feasible for a fixed range of uncertainty. In the second problem we will address the question of how to actually quantify flexibility.

15.2 Two-Stage Programming with Guaranteed Feasibility

In Chapter 14 we considered the model

$$
\begin{aligned}
\min \quad & f(d, z, \theta) \\
\text{s.t.} \quad & g(d, z, \theta) \leq 0,
\end{aligned}
\tag{15.1}
$$

where d is the vector of design variables (stage 1), z is the vector of control or recourse variables (stage 2), and θ is the vector of uncertain parameters. To account for the uncertainty of the parameters θ we considered the two-stage programming strategy. We indicated that one way to ensure that the stage-2 problem is feasible is to assign a penalty for the violation constraints as shown in (15.2),

$$
\min_{d} \; \mathop{E}_{\theta \in T} \left\{ \min_{z} C(d, z, \theta) + \sum_{j} c_j u_j \, \middle| \, f_j(d, z, \theta) \leq u_j, u_j \leq 0, j \in J \right\}.
\tag{15.2}
$$

However, let us assume we would like to enforce feasible operation for a specified uncertainty set T, for instance one that is given by lower and upper bounds of θ, that is for every $\theta \in T = \{\theta | \theta^L \leq \theta \leq \theta^U\}$.

We would then reformulate problem (15.2) as follows,

$$\min_{d} \; \underset{\theta \in T}{E} \left\{ \min_{z} C(d,z,\theta) \,\middle|\, f(d,z,\theta) \leq 0 \right\}$$
$$\text{s.t.} \quad \forall \theta \in T \; \exists z \; \left(\forall j \in J, f_j(d,z,\theta) \leq 0 \right), \tag{15.3}$$

in which we have added as a constraint the condition that ensures that, for the given design d, by proper adjustment of the controls z the inequalities $f_j(d,z,\theta) \leq 0$, $\forall j \in J$ are satisfied for $\forall \theta \in T$. Or, stated more precisely, for $\forall \theta \in T$ there exists a vector of control variables z that can satisfy $\forall j \in J$ the inequalities $f_j(d,z,\theta) \leq 0$. Needless to say the constraint in problem (15.3) is nontrivial as it effectively involves an infinite number of constraints as they have to hold for $\forall \theta \in T$. This class of problems is denoted as semi-infinite programming problems (Hettich and Kortanek, 1993).

Given the complexity of integrating the multiple integral for the expected value in problem (15.3), let us consider a discrete approximation in which we consider N points, θ^i, $i = 1, \dots, N$, with probability w_i for which z^i is selected. Problem (15.3) can then be reformulated as

$$\min_{d, z^1, z^2, \dots, z^N} \sum_{i=1}^{N} w_i c\left(d, z^i, \theta^i\right)$$
$$\text{s.t.} \quad \forall \theta \in T \; \exists z \; \left(\forall j \in J, f_j(d,z,\theta) \leq 0 \right). \tag{15.4}$$

If we remove the semi-infinite feasibility constraint, we can reduce problem (15.4) to the multiscenario problem,

$$\min_{d, z^1, z^2, \dots, z^N} \sum_{i=1}^{N} w_i C\left(d, z^i, \theta^i\right)$$
$$\text{s.t.} \quad f\left(d, z^i, \theta^i\right) \leq 0 \quad i = 1, \dots, N. \tag{15.5}$$

Figure 15.1 shows an interpretation of problem (15.5) as a multiscenario problem in which the design variables d are selected over the four scenarios that each involve uncertain parameters θ^i, $i = 1, \dots, 4$, with probabilities w_i, $i = 1, \dots, 4$, and for which corresponding control variables z^i, $i = 1, \dots, 4$, are selected. Here the probabilities w_i can be interpreted as lengths of time periods in the multiscenario problem.

Figure 15.1 Representation of the variables involved in the formulation (15.5).

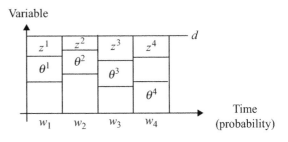

An algorithm for solving the two-stage programming problem (15.4) with the flexibility constraint is as follows.

(1) Set iteration counter $k = 0$, and select an initial set of parameters $T_0 = \{\theta^i, i = 1, 2, \ldots, N_0\}$.

(2) Solve the multiscenario problem (15.5) to determine the design variable d^k for the N_k scenarios.

(3) Apply the flexibility test for d^k.
   Check if $\forall \theta \in T \; \exists z \left(\forall j \in J, f_j(d, z, \theta) \leq 0 \right)$.
   (a) If the test is positive, stop.
   (b) If it is negative, identify the point θ^c with worst constraint violation.
      Define $T_{k+1} = T_k \cup \{\theta^c\} \quad N_{k+1} = |T_{k+1}|$.
      Set $k = k + 1$, and return to step (2).

We should note that the above algorithm corresponds to an iterative multiscenario problem in which the weights w_i are selected to approximate expected values. Furthermore, the set T_0 will normally contain the nominal value of the parameters, θ^N, that represents the expected value of θ. The remaining challenge is how to apply in step (3) the flexibility test for the fixed d^k. The answer to that question is provided in the next section.

15.3 Flexibility Analysis

As shown in Fig. 15.2, let us consider a fixed design d that is subjected to a vector of uncertain parameters θ, and for which there is a vector of control variables z that can be adjusted to counteract the effect of the uncertainties. Or, more specifically, our goal in the flexibility test is to determine if, for the given design d, $\forall \theta \in T$ a vector of control variables z can be selected to satisfy the inequalities $f_j(d, z, \theta) \leq 0 \; \forall j \in J$, where in the most common case $T = \{\theta | \theta^L \leq \theta \leq \theta^U\}$, where θ^L and θ^U are lower and upper bounds, respectively, for the uncertain parameters. Note that we might consider more general uncertainty sets such as $T = \{\theta | r(\theta) \leq 0\}$, where $r(\theta) \leq 0$ is a set of linear or nonlinear inequalities. Also, simplicial approximations might be used, as described in Goyal and Ierapetritou (2002).

As indicated above, the flexibility test can be symbolically expressed as follows:

$$\forall \theta \in T \left[\exists z \left(\forall j \in J, f_j(d, z, \theta) \leq 0 \right) \right]. \tag{15.6}$$

Figure 15.2 Elements of flexibility test for a fixed design d.

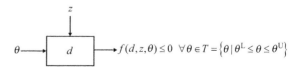

Equation (15.6) can, in turn, be shown to be equivalent to a constraint involving a max-min-max optimization problem as shown below (Halemane and Grossmann, 1983),

$$\forall \theta \in T \left[\exists z \left(\forall j \in J, f_j(d, z, \theta) \leq 0 \right) \right]$$

$$\Leftrightarrow \forall \theta \in T \left[\exists z \; \max_{j \in J} f_j(d, z, \theta) \right] \leq 0$$

$$\Leftarrow \forall \theta \in T \; \min_z \; \max_{j \in J} f_j(d, z, \theta) \leq 0 \tag{15.7}$$

$$\Leftrightarrow \max_{\theta \in T} \; \min_z \; \max_{j \in J} f_j(d, z, \theta) \leq 0.$$

Thus, we have proved the following theorem.

Theorem 15.1 *The flexibility test* $\forall \theta \in T \left[\exists z \left(\forall j \in J, f_j(d, z, \theta) \leq 0 \right) \right]$ *is equivalent to the inequality* $\max_{\theta \in T} \; \min_z \; \max_{j \in J} f_j(d, z, \theta) \leq 0.$

Given the max-min-max function, we can define the *flexibility function* $\chi(d)$ for a fixed design d, as follows:

$$\chi(d) = \max_{\theta \in T} \; \min_z \; \max_{j \in J} f_j(d, z, \theta). \tag{15.8}$$

In this way by calculating the function $\chi(d)$, we can arrive at either of the two following conclusions.

(a) If $\chi(d) \leq 0$ it implies that the design d is feasible $\forall \theta \in T$.
(b) If $\chi(d) > 0$ it implies that the design d is infeasible for *some* $\theta \in T$.

These, then, provide an answer to the steps (3a) and (3b) in the iterative multiscenario optimization algorithm presented in the previous section.

15.4 Flexibility Test with No Control Variables

To gain some insight into the geometric interpretation of the flexibility function, $\chi(d)$, consider the particular case when there are no control variables z, that is $\dim(z) = 0$. For this, problem (15.8) reduces to the following maximization problem:

$$\chi(d) = \max_{\theta \in T} \; \max_{j \in J} f_j(d, \theta). \tag{15.9}$$

Since the order of the max operators can be reversed, (15.9) is also equivalent to

$$\chi(d) = \max_{j \in J} \; \max_{\theta^j \in T} f_j(d, \theta^j), \tag{15.10}$$

in which we select the inequality $j \in J$ that leads to the largest value of the maximization of each constraint $f_j(d, \theta^j)$. Formulation (15.10) then suggests the following algorithm.

(1) Solve $u^j = \max_{\theta^j \in T} f_j(d, \theta^j) \quad \forall j \in J$.
(2) Set $\chi(d) = \max_{j \in J} \{u^j\}$.

Figure 15.3 Example of three inequalities in terms of a single uncertain parameter θ.

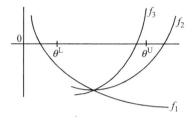

(3) If $\chi(d) \leq 0 \Rightarrow$ feasible $\forall \theta \in T$
$\chi(d) > 0 \Rightarrow$ infeasible for some $\theta \in T$.

As an example, consider the case of the three inequalities f_1, f_2, and f_3 in Fig.15.3 that involve one single uncertain parameter θ.

By maximizing each inequality with respect to the parameter θ, the algorithm yields the following results:

$$u_1 = \max_{\theta} f_1 < 0 \text{ at } \theta^{\mathrm{L}}$$
$$u_2 = \max_{\theta} f_2 < 0 \text{ at } \theta^{\mathrm{U}}$$
$$u_3 = \max_{\theta} f_3 > 0 \text{ at } \theta^{\mathrm{U}}.$$

From the above it is clear that the critical point θ^c is given by the upper bound, that is $\theta^c = \theta^{\mathrm{U}}$, since it is the point at which the maximum constraint violation takes place, namely $f_3(d, \theta^c) > 0$.

For the general case in which we have p uncertain parameters, θ_i, $i = 1, \ldots, p$, we should note that the full numerical maximization of each inequality $f_j(d, z, \theta)$, $j \in J$ corresponds to an NLP with respect to each parameter θ_i, $i = 1, \ldots, p$. These optimizations can be avoided if each inequality $f_j(d, z, \theta)$ is quasiconvex, as then either $\theta_i^c = \theta_i^{\mathrm{L}}$ or $\theta_i^c = \theta_i^{\mathrm{U}}$, meaning we need only to analyze the vertices (combinations of lower and upper bounds) in the parameter set $T = \{\theta | \theta^{\mathrm{L}} \leq \theta \leq \theta^{\mathrm{U}}\}$.

Furthermore, if $f_j(d, \theta)$ is monotone in each θ_i, only the sign of the gradients need to be analyzed, that is,

$$\frac{\partial f_j}{\partial \theta_i} > 0 \Rightarrow \theta_i^c = \theta_i^{\mathrm{U}}, \quad \frac{\partial f_j}{\partial \theta_i} < 0 \Rightarrow \theta_i^c = \theta_i^{\mathrm{L}}. \tag{15.11}$$

This can easily be verified with the example in Fig. 15.3.

As a final point, we should note that if, instead of the inequalities $f_j(d, z, \theta)$, $j \in J$, we are given the model in the form of equation and inequalities in terms of the state variables x and uncertain parameters θ, that is,

$$h(d, x_i, \theta) = 0$$
$$g(d, x, \theta) \leq 0, \tag{15.12}$$

then step (1) in the above algorithm involves solving the following LP or NLP:

$$u^j = \max_{\theta^j \in T} g_j(d, x, \theta^j)$$
$$\text{s.t.} \quad h(d, x, \theta^j) = 0. \tag{15.13}$$

15.5 Flexibility Test with Control Variables

For the general case when there are control variables z, that is $\dim(z) > 0$, consider the flexibility test in (15.8), that is,

$$\chi(d) = \max_{\theta \in T} \min_{z} \max_{j \in J} f_j(d, z, \theta). \tag{15.14}$$

To analyze (15.14) we define the inner minimax problem as the *feasibility function* $\psi(d, \theta)$ for fixed design d and fixed uncertain parameter θ,

$$\psi(d, \theta) = \min_{z} \max_{j \in J} f_j(d, z, \theta). \tag{15.15}$$

This feasibility function, which corresponds to a nondifferentiable optimization problem due to the max operator, is equivalent to the following optimization problem,

$$\begin{aligned} \psi(d, \theta) = \ & \min u \\ \text{s.t.} \quad & u \geq f_j(d, z, \theta) \quad j \in J \\ & u \in R^1, z \in R^{n_z}, \end{aligned} \tag{15.16}$$

which corresponds to an **LP** or **NLP** depending on whether the functions $f_j(d, z, \theta)$ are linear or nonlinear, and where u is a scalar variable that represents the maximum constraint violation that is being minimized. We should note that the function $\psi(d, \theta)$ represents a projection from the (d, z, θ) space into the (d, θ) space. With problem (15.16) the flexibility test (15.14) can be formulated as the following maximization problem:

$$\chi(d) = \max_{\theta \in T} \psi(d, \theta). \tag{15.17}$$

To provide a geometric interpretation of problems (15.16) and (15.17), consider the two following linear inequalities:

$$\begin{aligned} f_1 &= -z + \theta \leq 0 \\ f_2 &= z - 2\theta + 2 - d \leq 0. \end{aligned} \tag{15.18}$$

Let us assume that we set $d = 0.5$ and that we would like to determine whether the inequalities in (15.18) can be satisfied for all θ in the interval, $1 \leq \theta \leq 2$.

By plotting these inequalities in Fig. 15.4, it is clear that they are infeasible for the interval $1 \leq \theta \leq 1.5$, while they are feasible for the interval $1.5 \leq \theta \leq 2$, if z is adjusted for each θ. Furthermore, the critical point θ^c with largest violation of constraints corresponds to the lower bound $\theta^L = 1$.

These observations can in fact be verified from the formulations presented above. Consider the feasibility function in (15.16) applied to the inequalities in (15.18),

$$\begin{aligned} \psi(d, \theta) = \ & \min_{u, z} u \\ \text{s.t} \quad & u \geq -z + \theta \\ & u \geq z - 2\theta + 1.5. \end{aligned} \tag{15.19}$$

To solve problem (15.19), consider the Lagrange function, where λ_1, λ_2 are the nonnegative Lagrange multipliers of the two inequalities in (15.19),

Figure 15.4 Plot of feasible region for the inequalities in (15.18).

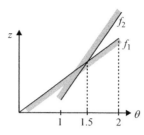

Figure 15.5 Plot of the function $\psi(d,\theta)$ in (15.22).

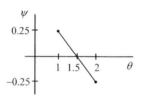

$$\mathcal{L} = u + \lambda_1(-z + \theta - u) + \lambda_2(z - 2\theta + 1.5 - u). \tag{15.20}$$

From the stationary conditions of the Lagrange function, we can conclude the following:

$$\frac{\partial \mathcal{L}}{\partial u} = 1 - \lambda_1 - \lambda_2 = 0 \Rightarrow \text{at least } \lambda_1 \text{ or } \lambda_2 > 0$$

$$\frac{\partial \mathcal{L}}{\partial z} = -\lambda_1 + \lambda_2 = 0 \Rightarrow \lambda_1 = \lambda_2 \text{ two active constraints} \tag{15.21}$$

$$\Rightarrow \lambda_1 = \lambda_2 = 0.5.$$

From (15.21) it follows that, since $\lambda_1 = \lambda_2 = 0.5$, the two inequalities are active. Furthermore, the Lagrange function in (15.20) at the optimum can be equated to the feasibility function $\psi(d,\theta)$,

$$\mathcal{L} = \psi(d,\theta) = 0.5\theta - \theta + 0.75 = 0.75 - 0.5\theta. \tag{15.22}$$

Figure 15.5 shows the plot of this function, indicating infeasibility for the interval $1 \le \theta < 1.5$ because $\psi > 0$, and feasibility for the interval $1.5 \le \theta \le 2$ because $\psi \le 0$. Furthermore, the critical point θ^c with largest value of $\psi = 0.25$ corresponds to the lower bound $\theta^L = 1$.

If we change the design variable d from 0.5 to 1, then, as can be seen in Fig. 15.6a, feasibility is achieved for all θ in the interval $1 \le \theta \le 2$, and $\psi \le 0$ as seen in the Fig. 15.6b, with the critical point θ^c corresponding to the lower bound $\theta^L = 1$ with $\psi = 0$.

Consider that we now add constraint $f_3 = -z + 6\theta - 9d \le 0$. In that case the feasible region is shown in Fig. 15.7a, where it can be seen that feasibility is also achieved for all θ in the interval $1 \le \theta \le 2$, but in this case ψ is a nondifferentiable piecewise linear function as

Figure 15.6 (a) Feasible region, (b) feasibility function ψ for $d = 1$.

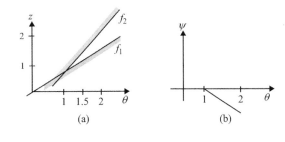

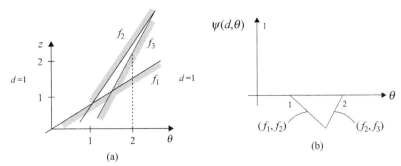

Figure 15.7 (a) Feasible region, (b) feasibility function for additional inequality f_3 for $d = 1$.

the left portion is given by the active constraints f_1, f_2, while the right portion is given by the active constraints f_2, f_3. Also, note that the critical points lie at both the lower and upper bounds, $\theta = 1$ and $\theta = 2$, because at these two points the feasibility function ψ attains its maximum value of $\psi = 0$. Thus, this simple example gives some insights into the nature of the feasibility function.

15.6 Parametric Region of Feasible Operation and Vertex Search

Having defined the feasibility function $\psi(d, \theta)$, we consider its projected feasible region $\mathcal{R}$ in the space of the uncertain parameters θ, namely,

$$\mathcal{R} = \left\{ \theta | \exists z \left(\forall j \in J, f_j(d, z, \theta) \leq 0 \right) \right\}$$
$$\Leftrightarrow \mathcal{R} = \left\{ \theta | \min_z \max_{j \in J} f_j(d, z, \theta) \right\} = \{\theta | \psi(d, \theta) \leq 0\}. \tag{15.23}$$

For the case of two uncertain parameters, θ_1, θ_2, we show in Fig. 15.8 the feasible region $\mathcal{R}$ in which it can be seen that the boundary is defined by the feasibility function $\psi(d, \theta)$. This means that the boundary is expressed in terms of the optimization problem in (15.16).

We can then also visualize the flexibility test, as shown in Figs. 15.9a and 15.9b.

As can be seen in Fig. 15.9a, feasibility is achieved for the specified uncertainty set T because $\psi(d, \theta) \leq 0$ for all $\theta \in T$, $\chi(d) \leq 0$. Also the critical point θ^c lies at the boundary

Figure 15.8 Parametric region $\mathcal{R}$ of feasible operation.

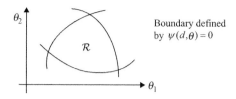

Figure 15.9 (a) Feasibility achieved for all $\theta \in T$, (b) infeasibility for some $\theta \in T$.

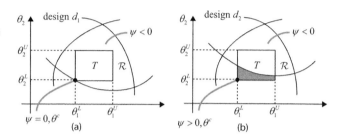

$\psi(d, \theta^c) = 0$ of the region $\mathcal{R}$ and corresponds to the vertex $\left(\theta_1^L, \theta_2^L\right)$ of the uncertainty set T. In contrast, Fig. 15.9b shows a region $\mathcal{R}$ that is infeasible for a subset of points in the set T. Furthermore, the maximum value of the feasibility function $\psi(d, \theta^c) > 0$ is also attained at the vertex $\left(\theta_1^L, \theta_2^L\right)$, and hence $\chi(d) > 0$.

We next describe an algorithm for solving the flexibility test problem for a fixed design d and a specified uncertainty set T,

$$\chi(d) = \max_{\theta \in T} \min_{z} \max_{j \in J} f_j(d, z, \theta). \tag{15.24}$$

Assuming that critical points θ^c correspond to vertices of the uncertainty set $T = \left\{\theta | \theta^c \leq \theta \leq \theta^U\right\}$, the max-min-max problem in (15.24) can be solved as follows.

(1) Define the vertices θ^k, $k \in V$, where $V = \{1, 2, \ldots, 2^p\}$ and $p = \dim(\theta)$.
(2) Solve, for each parameter vertex θ^k $k \in V$, the following optimization problem,

$$\psi\left(d, \theta^k\right) = \min u$$
$$\text{s.t.} \quad u \geq f_j\left(d, z, \theta^k\right), j \in J \tag{15.25}$$
$$u \in R^1, z \in R^{n_z},$$

where the scalar u represents the largest violation of the inequalities $f_j\left(d, z, \theta^k\right) \leq 0, j \in J$.
(3) Set $\chi(d) = \max_{k \in V}\left\{\psi\left(d, \theta^k\right)\right\}$.
(4) If $\chi(d) \leq 0$ the design d is guaranteed to be feasible for all points in the set T; if $\chi(d) > 0$ the design d is infeasible over the uncertainty set T.

The above algorithm requires, in step (2), the solution of 2^p optimization problems where $p = \dim(\theta)$. It is clear that if the number of parameters is large, this algorithm can be computationally expensive. Furthermore, we should note that if the region $\mathcal{R}$ is nonconvex, the critical points θ^c may not necessarily correspond to vertices (see Fig. 15.13).

15.7 Flexibility Index and Vertex Search

The flexibility test only provides a yes or no answer to the capability of a given design to achieve feasible operation over a specified parameter set T. Next we consider instead the problem of determining a measure of the size of the feasible region ($\mathcal{R}$) that, in turn, provides a guaranteed range for the uncertain parameters θ for which feasibility can be guaranteed by proper adjustment of the control variables z.

The measure of the region ($\mathcal{R}$), which we will denote as the *flexibility index* (Swaney and Grossmann, 1985a, b), is given with respect to a nominal parameter point θ^N, and expected parameter deviations $\Delta\theta^+$, $\Delta\theta^-$. If we then consider as a target for flexibility the set T defined by the nominal point and expected deviations,

$$T^* = \left\{\theta \mid \theta^N - \Delta\theta^- \leq \theta \leq \theta^N + \Delta\theta^+\right\}, \tag{15.26}$$

we can define, for the parametric set $T(F)$ in (15.27),

$$T(F) = \left\{\theta \mid \theta^N - F\Delta\theta^- \leq \theta \leq \theta^N + F\Delta\theta^+\right\}, \tag{15.27}$$

the largest scalar F such that $T(F)$ is inscribed within the region $\mathcal{R}$, as shown in Fig. 15.10. We denote the scalar F as the *flexibility index*. Once we obtain its value, we can conclude the following for a given design d depending on the value of the index:

$F = 1 \Rightarrow$ design meets the flexibility target,
$F > 1 \Rightarrow$ design exceeds the target,
$F < 1 \Rightarrow$ design does not meet the target.

To develop a mathematical formulation for finding the largest rectangle $T(F)$ within the region $\mathcal{R}$ centered at θ^N, and with sides proportional to $\Delta\theta^+$, $\Delta\theta^-$, we define the following parametric set $T(\delta)$ in terms of the nonnegative variable δ:

Figure 15.10 Geometric interpretation of set $T(F)$ with flexibility index F.

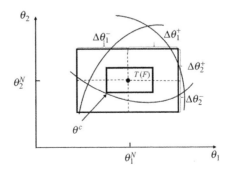

$$T(\delta) = \left\{ \theta | \theta^N - \delta \Delta \theta^- \leq \theta \leq \theta^N + \delta \Delta \theta^+ \right\}, \delta \geq 0. \tag{15.28}$$

The flexibility index can then be determined by selecting the largest value of δ such that the feasibility test holds true for all the parameter points θ in the set $T(\delta)$, that is,

$$F = \max \delta$$

$$\text{s.t.} \quad \max_{\theta \in T(\delta)} \min_z \max_{j \in J} f_j(d, z, \theta) \leq 0 \tag{15.29}$$

$$T(\delta) = \left\{ \theta | \theta^N - \delta \Delta \theta^- \leq \theta \leq \theta^N + \delta \Delta \theta^+ \right\}, \delta \geq 0.$$

Note from Fig. 15.10 that, at the optimum solution, $\psi(d, \theta^c) = 0$, meaning that the critical point θ^c lies at the boundary of the region $\mathcal{R}$, and limits the size of the largest rectangle.

The solution of problem (15.29) can be simplified if it is assumed critical points θ^c correspond to vertices of the parametric set $T(\delta)$. The basic idea could then be to find the largest deviation δ^k along each vertex direction $\tilde{\theta}^k$, $k \in V = \{1, 2, \ldots, 2^p\}$, and pick among these the smallest one to define the flexibility index. As an example, the vertex direction $\tilde{\theta}^1$ defined by positive directions is given by

$$\tilde{\theta}^1 = \begin{bmatrix} \Delta \theta_1^+ \\ \Delta \theta_2^+ \end{bmatrix}. \tag{15.30}$$

The algorithm for vertex search is then as follows.

(1) For each $k \in V = \{1, 2, \ldots, 2^p\}$, solve the following problem,

$$\delta^k = \max \delta$$

$$\text{s.t.} \quad f(d, z, \theta) \leq 0 \quad (\mathbf{P}^k)$$

$$\theta = \theta^N + \delta \Delta \tilde{\theta}^k, \quad \delta \geq 0.$$

(2) Set $F = \min_{k \in V} \left\{ \delta^k \right\}$.

We should note that having obtained the flexibility index we can determine the actual feasible range given by $\theta^N - F \Delta \theta^- \leq \theta \leq \theta^N + F \Delta \theta^k$. Furthermore, the critical point θ^c, corresponds to the vertex direction with smallest δ^k. As in the case of the feasibility test, the major limitation of the vertex search is that it requires the solution of 2^p optimization problems in step (2). The next section (Section 15.8) will show how to overcome this problem.

An interesting additional point is that, from the results of the above algorithm, we can perform a sensitivity analysis of the flexibility index with respect to changes in a given design variable d_i. For this, consider the variation of the flexibility index F with respect to the design variable d_i,

$$\frac{\partial F}{\partial d_i} = \sum_{j \in J} \frac{\partial F}{\partial f_j} \frac{\partial f_j}{\partial d_i}, \tag{15.31}$$

where $\frac{\partial f_j}{\partial d_i}$ can be obtained analytically or by finite differences from the inequalities $f_j(d, z, \theta)$.

Say $k = c$ corresponds to the closest vertex direction with deviation δ^c. Then,

$$\frac{\partial F}{\partial f_j} = \frac{\partial \delta^c}{\partial f_j} = -\lambda_j^c, \tag{15.32}$$

where λ_j^c corresponds to the Lagrange multiplier of the jth inequality in problem (P^k). Substituting in (15.31), one can then obtain an explicit equation to predict the sensitivity of the flexibility index with respect to a change in the design variable d_i.

$$\frac{\partial F}{\partial d_i} = -\sum_{j \in J} \lambda_j^c \frac{\partial f_j}{\partial d_i}. \tag{15.33}$$

This sensitivity could be very useful for retrofit design problems (e.g., to determine in a heat-exchanger network that exchanger where adding area leads to the largest increase in the flexibility index). This analysis can also be extended to multiobjective optimization problems in which cost is minimized and flexibility is maximized (see Pistikopoulos and Grossmann, 1988).

15.8 Theoretical Conditions for Vertex Solutions

As was shown in Sections 15.6 and 15.7, the flexibility test and calculation of the flexibility index can be greatly simplified if it is assumed that critical points correspond to vertices in the sets T and $T(\delta)$. An important theoretical question is to determine the sufficient conditions that need to hold for a critical point θ^c to be a vertex in these parameter sets. For simplicity, we restrict the presentation to the parameter set $T = \{\theta | \theta^L \leq \theta \leq \theta^U\}$.

A region $\mathcal{R} = \{\theta | \psi(d, \theta) \leq 0\}$ is defined as being one-dimensional (1-d) convex if and only if, for θ^1, $\theta^2 \in \mathcal{R}$, where $\theta^2 = \theta^1 + \beta e_j$, ($\beta \neq 0$, e_j coordinate direction), the point $\theta = \alpha \theta^1 + (1 - \alpha)\theta^2 \in \mathcal{R} \; \forall \alpha \in [0, 1]$. In short, 1-d convexity requires that the convexity condition holds true only along the coordinate directions.

Figure 15.11 shows examples of three regions. Figure 15.11a corresponds to a convex region, which is shown to be also 1-d convex. Figure 15.11b corresponds to a nonconvex region, which, however, is 1-d convex. Finally, Fig. 15.11c corresponds to a nonconvex

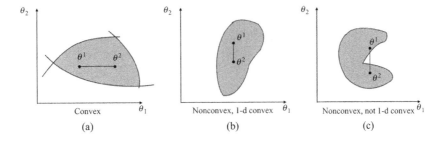

Figure 15.11 (a) Convex region, (b) 1-d convex region, (c) nonconvex region.

region, which is not 1-d convex. In summary, 1-d convexity can be viewed as a less restrictive condition than convexity.

The following theorems can be established for 1-dimensional convexity (Swaney and Grossmann, 1985a, b).

Theorem 15.2 *If $\psi(d,\theta)$ is a 1-d quasiconvex function in θ, then $\mathcal{R} = \{\theta | \psi(d,\theta) \leq 0\}$ is a 1-d convex region.*

Theorem 15.3 *If $\psi(d,\theta)$ is 1-d quasiconvex, then the critical point θ^c lies at a vertex of the set T.*

The above theorems rely on the following definition.

Definition 15.1 $\psi(d,\theta)$ is 1-d quasiconvex in θ if and only if
$$\psi(d, \alpha\theta^1 + (1-\alpha)\theta^2) \leq \max\{\psi(d,\theta^1), \psi(d,\theta^2)\} \quad \forall \alpha \in [0,1], \text{ where } \theta^1, \theta^2 \in \mathcal{R},$$
$$\theta^2 = \theta^1 + \beta e_j.$$

Figure 15.12 illustrates an example of a function $\psi(d,\theta)$ which is quasiconvex, while being a nonconvex concave function.

The following theorem provides a sufficient condition for the feasibility function $\psi(d,\theta)$ to be 1-d quasiconvex in θ.

Theorem 15.4 *If the constraint functions $f_j(d,z,\theta) \leq 0, j \in J$ are quasiconvex in z and 1-d quasiconvex in θ, then $\psi(d,\theta)$ is 1-d quasiconvex in θ.*

In summary, from the above theorems, the following are sufficient conditions for a critical point θ^c to lie at the vertex of the parameter set T.

(1) The functions $f_j(d,z,\theta) \leq 0, j \in J$ are quasiconvex in z and 1-d quasiconvex in θ.
(2) $\psi(d,\theta)$ is 1-d quasiconvex in θ.
(3) $\mathcal{R}$ is 1-d convex.

An example of a problem where the critical point does not correspond to a vertex (Biegler et al., 1997) is the heat-exchanger network shown in Fig. 15.13 where the heat capacity flowrate F_{H1} is an uncertain parameter. We would like to determine whether this network is feasible for the range $1 \leq F_{H1} \leq 1.8$ (kW/K).

The following inequalities are considered for feasible operation of this network:

Figure 15.12 Concave function $\psi(d,\theta)$ which is quasiconvex.

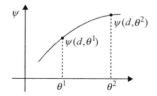

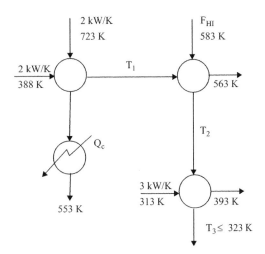

Figure 15.13 Heat-exchanger network with uncertain heat capacity flowrate, F_{H1}.

Feasibility in exchanger 2:	$T_2 - T_1 \geq 0$	
Feasibility in exchanger 3:	$T_2 - 393 \geq 0$	
Feasibility in exchanger 3:	$T_3 - 313 \geq 0$	(15.34)
Specification in outlet temperature:	$T_3 \leq 323.$	

By considering the corresponding heat balances, we can solve for the above temperatures in terms of the cooling load Q_c, the control variable, and in terms of F_{H1}, the uncertain parameter. The reduced inequalities in (15.34) are then as follows:

$$f_1 = -25 + Q_c[(1/F_{H1}) - 0.5] + 10/F_{H1} \leq 0$$
$$f_2 = -190 + (10/F_{H1}) + (Q_c/F_{H1}) \leq 0$$
$$f_3 = -270 + (250/F_{H1}) + (Q_c/F_{H1}) \leq 0 \tag{15.35}$$
$$f_4 = 260 - (250/F_{H1}) - (Q_c/F_{H1}) \leq 0.$$

If we now examine the two extreme points, for F_{H1}, by solving the NLP in (15.16) for the above inequalities we obtain the following.

(a) For $F_{H1} = 1$ kW/K, $\psi_1(1) = -5$, $Qc = 15$ kW.
(b) For $F_{H1} = 1.8$ kW/K, $\psi_2(1.8) = -5$, $Qc = 227$ kW.

Because $\psi_1 < 0$ and $\psi_2 < 0$, we may be tempted to conclude that the network is feasible to operate for the range $1 \leq F_{H1} \leq 1.8$ kW/K. However, let us consider an intermediate value, say $F_{H1} = 1.2$ kW/K, for problem (15.16). We then obtain:

$$F_{H1} = 1.2 \, \text{kW/K}, \psi(1.2) = 2.85; Qc = 58.57 \, \text{kW}.$$

In other words, the network is infeasible at the *nonvertex* point $F_{H1} = 1.5$ kW/K. If we plot the constraints in (15.34) as shown in Fig. 15.14 we can clearly see that we have a nonconvex region where for $1.118 \leq F_{H1} \leq 1.65$ we have infeasible operation. In fact, at

Figure 15.14 Feasible region for constraints in (15.34).

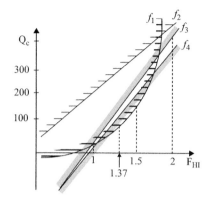

$F_{H1} = 1.37$ kW/K we have the greatest violation of constraints since at that point $\psi(1.37) = +5.108$ attains its maximum value. Hence, $F_{H1} = 1.37$ corresponds to the critical point.

The above example then shows that it is possible to have nonvertex critical points and, consequently, we need an appropriate method that will be able to predict such points, as we will show in the next section.

15.9 Active-Set Strategy

In this section we present mixed-integer programming formulations for the flexibility test (15.24) and flexibility index (15.29) that do not rely on the assumption that a critical point θ^c corresponds to a vertex. The basic idea in these formulations relies on performing the search in the space of active inequality constraints (Grossmann and Floudas, 1987).

The flexibility-test problem, $\chi(d, \theta) = \max_\theta \min_z \max_{j \in J} f_j(d, z, \theta)$ in (15.24), can be reformulated as the two-level optimization problem

$$\chi(d) = \max_{\theta \in T} \psi(d, \theta)$$
$$\text{s.t.} \quad \psi(d, \theta) = \min_z \max_{j \in J} f_j(d, z, \theta). \tag{15.36}$$

In order to replace the lower level optimization problem with a set of constraints, we consider the KKT conditions (Bazaraa et al., 2006) of the feasibility-test function $\psi(d, \theta)$,

$$\psi(d, \theta) = \min_z u$$
$$\text{s.t.} \quad f_j(d, z, \theta) - u \leq 0, \quad j \in J \tag{15.37}$$
$$u \in R^1,$$

where we assume that the functions $f_j(d, z, \theta)$ include the bounds for the control variables z. Therefore, $n = \dim(z) < m = |J|$.

In order to derive the KKT conditions for problem (15.36), we define the Lagrangean function

$$\mathcal{L} = u + \sum_{j \in J} \lambda_j \left[f_j(d, z, \theta) - u \right]. \tag{15.38}$$

The stationary conditions with respect to the scalar variable u and the control variables z are given by (15.39a) and (15.39b), while the complementarity conditions are given by (15.39c), where $\lambda_j \geq 0$ are the Lagrange multipliers,

$$\frac{\partial \mathcal{L}}{\partial u} = 1 - \sum_{j \in J} \lambda_j = 0 \qquad \text{(a)}$$

$$\frac{\partial \mathcal{L}}{\partial z} = \sum_{j \in J} \lambda_j \frac{\partial f_j}{\partial z} = 0 \qquad \text{(b)} \qquad\qquad (15.39)$$

$$\left.\begin{array}{l} \lambda_j \left[f_j(d, z, \theta) - u \right] = 0 \\[2mm] \lambda_j \geq 0 \quad f_j(d, z, \theta) - u \leq 0 \end{array}\right\} j \in J \quad \text{(c)}.$$

From (15.39a) we can conclude that at least one multiplier is strictly positive $\lambda_j > 0$ (i.e., one f_j is active). From (15.39b), if the transpose of the Jacobian, $J_z^T = \left[\frac{\partial f_1}{\partial z} \ \frac{\partial f_2}{\partial z} \cdots \frac{\partial f_m}{\partial z} \right]$ is full rank, then at least $n + 1$ multipliers must be strictly positive $\lambda_j > 0$. But from (15.39c), to have a solution $f_j(d, z, \theta) - u \leq 0$, at most $n + 1$ constraints are active ($n = \dim(z)$, $\dim(u) = 1$). This property, known as the Haar condition (Haar, 1918), holds for the minimax problem in (15.37). We can then establish the following useful property.

Property 15.1 If $\psi(d, \theta)$ has $n + 1$ active constraints, then the KKT are necessary and sufficient for a local min of $\psi(d, \theta)$.

The implication of this property is that we can use the KKT conditions to replace $\psi(d, \theta)$ as a constraint in problem (15.36), and then in turn identify the sets of active constraints as shown in the example in Fig. 15.15.

By replacing the KKT conditions for $\psi(d, \theta)$ in problem (15.36), the active-set formulation for the feasibility-test problem is as follows:

Figure 15.15 Example of active sets for three inequalities.

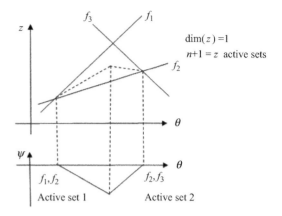

Nondifferentiable
maximum at bounds

$$\chi(d) = \max_{\theta \in T} \psi(d, \theta)$$

$$\text{s.t.} \quad 1 - \sum_{j \in J} \lambda_j = 0$$

$$\sum_{j \in J} \lambda_j \frac{\partial f_j}{\partial z} = 0 \tag{15.40}$$

$$\lambda_j \Big[f_j(d, z, \theta) - u \Big] = 0, \quad j \in J$$

$$\lambda_j \geq 0, f_j(d, z, \theta) - u \leq 0, \quad j \in J.$$

The difficulty, however, in problem (15.40) lies in the complementarity conditions as they involve bilinearities that, aside from introducing nonconvexities, give rise to discontinuities in the derivatives. To circumvent these difficulties, we explicitly model the discrete choice of active constraints involved in the complementarity conditions. First, we define the slack variables, $s_j \geq 0$, such that the inequalities are expressed as follows:

$$s_j = u - f_j(d, z, \theta) \; \forall j \in J. \tag{15.41}$$

Next, we introduce the binary variables y_j to indicate which constraints are active or not, namely,

$$y_j = \begin{cases} 1 \text{ if constraint } j \text{ is active } (s_j = 0) \\ 0 \text{ otherwise } (s_j \geq 0). \end{cases} \tag{15.42}$$

The complementarity conditions, together with the Haar condition, can then be expressed as follows:

$$\left. \begin{array}{ll} s_j - U\big(1 - y_j\big) \leq 0, & j \in J \\ \lambda_j - y_j \leq 0 & j \in J \end{array} \right\}$$

$$\sum_{j \in J} y_j \leq n + 1 \tag{15.43}$$

$$s_j, \lambda_j \geq 0 \quad j \in J.$$

By substituting (15.41) and (15.43) into (15.40) the flexibility test can then be formulated as the following mixed-integer programming problem (PFT),

$$\chi(d) = \max_{\theta, u, z, \lambda_j, s_j, y_j} u$$

$$\text{s.t.} \quad 1 - \sum_{j \in J} \lambda_j = 0$$

$$\sum_{j \in J} \lambda_j \frac{\partial f_j}{\partial z} = 0$$

$$s_j + f_j(d, z, \theta) = u \qquad j \in J \tag{PFT}$$

$$s_j - U\left(1 - y_j\right) \le 0, \quad j \in J$$

$$\lambda_j - y_j \le 0, \qquad\qquad j \in J$$

$$\sum_{j \in J} y_j \le n + 1$$

$$y_j \in \{0, 1\}, \lambda_j, s_j \ge 0, u \in R^1, z \in R^n, \theta^{\mathrm{L}} \le \theta \le \theta^{\mathrm{U}}.$$

Depending on the nature of the inequalities problem, (PFT) corresponds to an MILP or an MINLP problem. In particular, if the functions $f_j(d, z, \theta)$ are linear in z and θ then the gradients $\frac{\partial f_j}{\partial z}$ are constants with which (PFT) reduces to an MILP problem. Furthermore, the KKT conditions are necessary and sufficient, and the maximum number of active constraints is given by $\sum_{j \in J} y_j \le n + 1$. In the case that the functions $f_j(d, z, \theta)$ are nonlinear in z, θ, problem (PFT) gives rise to an MINLP problem. For the particular case that the constraints $f_j(d, z, \theta)$ are quasiconvex in z and quasiconcave in θ, the relaxation of the MINLP will have a unique solution with which convex MINLP solvers can be used. Interestingly, that is the case of the heat-exchanger network problem in Fig. 15.13 where (PFT) can predict the nonvertex critical point $F_{\mathrm{H1}} = 1.37$ (Biegler et al., 1997). For the case that the constraints $f_j(d, z, \theta)$ are quasiconvex in z and 1-d quasiconvex in θ, (PFT) will correspond to a nonconvex MINLP that yields vertex solutions but without requiring the solution of 2^p subproblems, as is the case with vertex search methods. The computational effort of the mixed-integer programming problem will be proportional to the number of constraints, $|J|$. Finally, we should also note that in problem (PFT) the bounds on the uncertain parameters, $\theta^{\mathrm{L}} \le \theta \le \theta^{\mathrm{U}}$, can easily be replaced by more general parameter sets $r(\theta) \le 0$.

Another important point about problem (PFT) is that it can easily be extended for the case where the original model is given by equations and inequalities as in (15.12). By assigning multipliers to these constraints as in (15.44),

$$\begin{aligned} h_i(d, x, z, \theta) &= 0, \quad i \in I \to \mu_i \\ g_j(d, x, z, \theta) &\le 0, \quad j \in J \to \lambda_i, \end{aligned} \tag{15.44}$$

and defining the feasibility-test problem as follows,

$$\psi(d,\theta) = \min_z u$$
$$h_i(d,x,z,\theta) = 0, \quad i \in I \qquad (15.45)$$
$$g_j(d,x,z,\theta) \le u, \quad j \in J.$$

The derivation then follows similar steps as in the derivation of problem (PFT) (see Exercise 15.2).

As for the flexibility index, we can also formulate it as a mixed-integer programming problem in terms of active sets, by defining the index F as the smallest deviation δ to the boundary $\psi(d,\theta) = 0$. That is,

$$F = \min \delta$$
$$\text{s.t.} \quad \psi(d,\theta) = 0 \qquad (15.46)$$
$$T(\delta) = \{\theta | \theta^N - \delta\Delta\theta^- \le \theta \le \theta^N + \delta\Delta\theta^+\}.$$

By replacing $\psi(d,\theta)$ by the KKT conditions, and by introducing the binary variables y_j, the flexibility index can be formulated as the following mixed-integer programming problem,

$$F = \min_{\theta,\delta,z,\lambda_j,s_j,y_j} \delta$$

$$\text{s.t.} \quad 1 - \sum_{j \in J} \lambda_j = 0$$

$$\sum_{j \in J} \lambda_j \frac{\partial f_j(d,z,\theta)}{\partial z} = 0$$

$$s_j + f_j(d,z,\theta) = 0 \qquad j \in J$$

$$s_j - U\left(1 - y_j\right) \le 0 \quad j \in J \qquad \text{(PFI)}$$

$$\lambda_j - y_j \le 0 \qquad\qquad j \in J$$

$$\sum_{j \in J} y_j \le n + 1 \qquad j \in J$$

$$\theta^N - \delta\Delta\theta^- \le \theta \le \theta^N + \delta\Delta\theta^+$$

$$y_j \in \{0,1\}, \quad \lambda_j, s_j \ge 0, \quad u \in \mathbb{R}^1, \delta \ge 0, \quad z \in \mathbb{R}^n$$

$$\theta^L \le \theta \le \theta^U.$$

As in the case of problem (PFT), (PFI) is a mixed-integer program, which corresponds to an MILP or MINLP depending on the nature of the constraint functions, and is not restricted to assuming vertex solutions. Also, this formulation can be extended to handle models with equality and inequality constraints as in (15.22).

For the case when $f_j(d,z,\theta)$ is monotone in z and the Jacobian J_z^T is full rank, the set of $n+1$ active constraints can be identified from the stationary condition in terms of the control variable z.

Figure 15.16 Example with three inequalities.

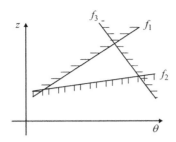

As an example, consider the inequalities in Fig. 15.16. It is clear from

$$\lambda_1 \frac{\partial f_1}{\partial z} + \lambda_2 \frac{\partial f_2}{\partial z} + \lambda_3 \frac{\partial f_3}{\partial z} = 0 \qquad (15.47)$$

that, because $\frac{\partial f_1}{\partial z} > 0$, $\frac{\partial f_2}{\partial z} < 0$, $\frac{\partial f_3}{\partial z} > 0$, and that there are $n + 1 = 2$ active constraints, the two sets of active constraints are given by f_1, f_2 and f_2, f_3, since these satisfy (15.47) for nonnegative multipliers. Identifying these two active sets, we can simply fix their corresponding binary variables in the MI(N)LP either (PFT) or (PFI), and obtain the solution by solving the two corresponding LPs or NLPs.

EXERCISES

15.1 The inequality constraints for feasible operation of a design d are given by

$$f_1 = -25\theta + z[1 - \theta/2] + d \leq 0$$
$$f_2 = -190\theta + z + d \leq 0$$
$$f_3 = \quad 260\theta - z - 240 - d \leq 0,$$

where θ is an uncertain parameter and z is a control variable. For the design $d = 10$, do the following.
(a) Plot the feasible region of operation in the $z - \theta$ space.
(b) Obtain the analytical expression for the feasibility function $\psi(d, \theta)$ in the range $0.5 \leq \theta \leq 2$, and plot this function.
(c) Determine the critical point for feasible operation in this design. Explain why the critical point is a vertex or a nonvertex solution.
(d) Is this design feasible for the parameter range $0.5 \leq \theta \leq 2$?

15.2 Derive the mathematical formulations for the active-set strategy for the following cases.
(a) Feasibility test: only inequalities, no control variables.
(b) Feasibility test: equalities and inequalities with control variables.
(c) Flexibility index for two cases above.

15.3 Given is a set of inequalities $f_j(d, z, \theta, \sigma) \leq 0, j \in J$ that are linear in d, z, θ, and σ, and where d is a fixed design, z are the control variables, and θ and σ are uncertain parameters. The difference between parameters θ and σ is that for θ we can manipulate the variables z to meet the constraints, while σ are parameters for which this is not possible because we cannot readily measure and react to changes to this parameter. An example for θ would be inlet flow that can be readily measured. An example for σ would be a tray efficiency that cannot be readily measured. Assuming that θ and σ are specified in the bounded sets $T = \{\theta|\theta^L \leq \theta \leq \theta^U\}$, $S = \{\sigma|\sigma^L \leq \sigma \leq \sigma^U\}$, do the following.

(a) Formulate a mathematical expression in terms of max and min operators to impose the flexibility condition that the inequalities must be satisfied $\forall \theta \in T$, $\forall \sigma \in S$.

(b) Derive an MILP formulation to solve the above flexibility test problem.

15.4 In the heat-exchanger network shown below, the inlet temperatures of the two hot and two cold process streams are regarded as uncertain parameters. Given the nominal values of the temperatures shown in the figure, and expected deviations of ± 10 K in each of these streams, determine the flexibility index for this network and its range of inlet temperatures for feasible operation.

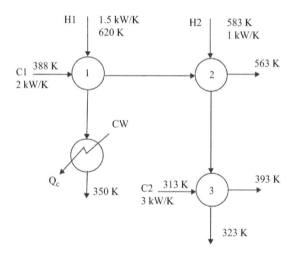

You can do the following to solve this problem.

(a) Formulate the inequality constraints for feasible heat exchange and the spec $T \leq 323$ K in terms of the cooling load Q_c and the inlet temperatures using $\Delta T_{\min} = 0$ K.

(b) Solve for the flexibility index with a vertex enumeration scheme (i.e., sixteen LPs) and with the MILP formulation.

Note: Areas are not specified. Q_c at 300 K.

15.5 Derive the mixed-integer active-set strategy formulation of the flexibility index for the case when the model for a fixed design is given by equations and inequalities with state variables x, with more than one control variable $z(n_z \geq 1)$, and with uncertain parameters θ.

APPENDIX A

Modeling Systems and Optimization Software

A.1 Modeling Systems

Review: Kallrath, J. (2004) *Modeling Languages in Mathematical Optimization*, Kluwer Academic Publishers.

AIMMS www.aimms.com/
AIMMS (Advanced Interactive Multidimensional Modeling System) is a prescriptive analytics software. It has two main product offerings that provide modeling and optimization capabilities across a variety of industries. The AIMMS Prescriptive Analytics Platform allows advanced users to develop optimization-based applications and deploy them to business users. AIMMS SC Navigator, launched in 2017, is built on the AIMMS Prescriptive Analytics Platform and provides configurable Apps for supply-chain teams. SC Navigator provides supply-chain analytics to nonadvanced users.

Bisschop, J. and Roelofs, M. (2004) "The modeling language AIMMS", in Kallrath, J. (ed.) *Modeling Languages in Mathematical Optimization*. Applied Optimization, vol. 88. Boston, MA: Springer.

AMPL https://ampl.com/
The AMPL system is a modeling tool that supports the entire optimization modeling lifecycle: development, testing, deployment, and maintenance. By using a high-level representation that represents optimization models, AMPL promotes rapid development and reliable results. AMPL integrates a modeling language for describing optimization data, variables, objectives, and constraints; a command language for browsing models and analyzing results; and a scripting language for gathering and manipulating data and for implementing iterative optimization schemes. All use the same concepts and syntax for streamlined application-building.

Fourer, R. (2013) "Algebraic modeling languages for optimization", in Gass, S. I. and Fu, M.C. (eds.) *Encyclopedia of Operations Research and Management Science*, Springer.

FICO XPRESS www.fico.com/en/products/fico-xpress-optimization
The FICO XPRESS optimizer is a commercial optimization solver for linear programming (LP), mixed-integer linear programming (MILP), convex quadratic programming (QP), convex quadratically constrained quadratic programming (QCQP), second-order cone programming (SOCP), and their mixed-integer counterparts. XPRESS includes a general purpose nonlinear solver, XPRESS NonLinear, including a successive linear programming

algorithm (SLP, first-order method), and Artelys Knitro (second-order methods). XPRESS was originally developed by Dash Optimization, and was acquired by FICO in 2008.

Ashford, R. (2007) "Mixed integer programming: A historical perspective with Xpress-MP", *Annals of Operations Research*, **149**, 5–17.

GAMS www.gams.com/

GAMS is one of the leading tool providers for the optimization industry and was the first software system to combine the language of mathematical algebra with traditional programming concepts in order to efficiently describe and solve optimization problems. The General Algebraic Modeling System (GAMS) is a high-level modeling system for mathematical programming and optimization. It consists of a language compiler and a stable of integrated high-performance solvers. GAMS is tailored for complex, large-scale modeling applications, and allows us to build large maintainable models that can be adapted quickly to new situations. GAMS is specifically designed for modeling linear, nonlinear, and mixed-integer optimization problems.

Bussieck, M.R. and Meeraus, A. (2004) "General Algebraic Modeling System (GAMS)", in Kallrath, J. (ed.) *Modeling Languages in Mathematical Optimization*, Applied Optimization, vol. 88. Boston, MA: Springer, pp. 137–157.

Julia https://julialang.org/

Julia is a high-level, high-performance dynamic programming language for technical computing. It is free (open source) and supports Windows, OSX, and Linux. It has a familiar syntax, works well with external libraries, is fast, and has advanced language features like metaprogramming that enable interesting possibilities for optimization software.

Bezanson, J. et al. (2017) "Julia: A fresh approach to numerical computing", *SIAM Review*, **59**, 65–98.

JuMP https://jump.dev/JuMP.jl/stable/

JuMP is a domain-specific modeling language for mathematical optimization embedded in Julia. It currently supports a number of open-source and commercial solvers for a variety of problem classes, including linear programming, mixed-integer programming, second-order conic programming, semidefinite programming, and nonlinear programming.

Dunning, I., Huchette, J. and Lubin, M. (2017) "JuMP: A modeling language for mathematical optimization", *SIAM Review*, **59**(2), 295–320.

LINDO Systems www.lindo.com/

LINGO is a comprehensive tool designed to make building and solving linear, nonlinear (convex and nonconvex/global), quadratic, quadratically constrained, second-order cone, semidefinite, stochastic, and integer optimization models faster, easier, and more efficiently. LINGO provides an integrated package that includes a language for expressing optimization models, a full featured environment for building and editing problems, and a set of fast built-in solvers.

Cunningham, K. and Schrage, L. (2004) "The LINGO algebraic modeling language", in Kallrath, J. (ed.) *Modeling Languages in Mathematical Optimization*. Applied Optimization, vol. 88. Boston, MA: Springer.

Pyomo www.pyomo.org/

Pyomo was conceived in the mid 2000s at Sandia National Labs, implementing custom representations, algorithms, and decompositions for large-scale models. Object-oriented modeling and the natural expression of model concepts are the guiding principles behind Pyomo, which embeds its modeling objects in Python, granting modelers and developers easy interoperability to a rich set of extensions and supporting libraries. Traditional algebraic modeling languages provide a high-level abstraction between algebraic formulations and the matrix input to optimization solvers. Pyomo extends these abstractions to support stochastic programming (PySP), differential equations (Pyomo.DAE), and propositional logic (Pyomo.GDP). At the same time, it provides automated access to low-level operations on model expressions, such as symbolic differentiation and bounds tightening. Pyomo can interface to GAMS and AMPL-linked free/commercial solvers, the Z3 satisfiability solver, or custom algorithms implemented in Pyomo supplies the backbone modeling capabilities behind the IDAES PSE framework, Egret, and other higher-level engineering analysis tools.

Hart, W. E. et al. (2017) *Pyomo – Optimization Modeling in Python*. Springer.

A.2   Optimization Software: LP, MILP, NLP, MINLP, GDP

A.2.1   Optimization Software Reviews

Linear Programming Software Survey, R. Fourer
www.informs.org/ORMS-Today/Public-Articles/June-Volume-40-Number-3/Linear-Programming-Software-Survey

Nonlinear Programming Software Survey, S. Nash
www.informs.org/ORMS-Today/OR-MS-Today-Software-Surveys/Nonlinear-Programming-Software-Survey

Atamturk, A. and Savelsbergh, M. W. P. (2005) "Integer-programming software systems", *Annals of Operations Research*, **140**, 67–124.

D'Ambrosio, C. and Lodi, A. (2013) "Mixed integer nonlinear programming tools: an updated practical overview", *Annals of Operations Research*, **204**(1), 301–320.

Gould, N., Orban, D. and Toint, P. (2005) "Numerical methods for large-scale nonlinear optimization", *Acta Numerica*, 299–361.

Kronqvist, J. et al. (2019) "A review and comparison of solvers for convex MINLP", *Optimization and Engineering*, **20**, 397–455.

Kumar, P. H. and Mageshvaran, R. (2020) "Methods and solvers used for solving mixed integer linear programming and mixed nonlinear programming problems: A review", *International Journal of Scientific & Technology Research*, **9**, 1872–1882.

Linderoth, J. T. and Ralphs, T. K. (2005) "Noncommercial software for mixed-integer linear programming". http://homepages.cae.wisc.edu/~linderot/papers/Linderoth-Ralphs-05-TR.pdf

A.2.2 Solvers and Corresponding Links

LP

- CLP https://github.com/coin-or/Clp
- CPLEX www.ibm.com/analytics/cplex-optimizer
- GLPK www.gnu.org/software/glpk/
- GUROBI www.gurobi.com/
- MOSEK www.mosek.com/
- QSOPT www.math.uwaterloo.ca/~bico/qsopt/
- SoPlex https://soplex.zib.de/
- XA www.maximalsoftware.com/solvers/xa.html
- XPRESS www.fico.com/en/products/fico-xpress-optimization

MILP

- CBC https://github.com/coin-or/Cbc
- CPLEX www.ibm.com/analytics/cplex-optimizer
- GLPK www.gnu.org/software/glpk/
- GUROBI www.gurobi.com/
- MOSEK www.mosek.com/
- SCIP www.scipopt.org/
- XA www.maximalsoftware.com/solvers/xa.html
- XPRESS www.fico.com/en/products/fico-xpress-optimization

MIQP

- CPLEX www.ibm.com/analytics/cplex-optimizer
- GUROBI www.gurobi.com/
- XPRESS www.fi co.com/en/products/fi co-xpress-optimization

NLP

- Antigone http://ares.tamu.edu/ANTIGONE/
- BARON www.minlp.com/baron
- CONOPT www.conopt.com/
- COUENNE https://github.com/coin-or/Couenne
- IPOPT https://github.com/coin-or/Ipopt
- KNITRO www.artelys.com/solvers/knitro/

- LANCELOT www.swmath.org/software/500
- LGO www.gams.com/latest/docs/S_LGO.html
- LINDO www.lindo.com/index.php/products/lingo-and-optimization-modeling
- LOQO https://vanderbei.princeton.edu/loqo/LOQO.html
- MINOS www.sbsi-sol-optimize.com/asp/sol_product_minos.htm
- SNOPT www.sbsi-sol-optimize.com/asp/sol_product_snopt.htm

MINLP

- Alpha-ECP www.sbsi-sol-optimize.com/asp/sol_product_snopt.htm
- Antigone http://ares.tamu.edu/ANTIGONE/
- BARON www.minlp.com/baron
- Bonmin https://github.com/coin-or/Bonmin
- COUENNE https://github.com/coin-or/Couenne
- DICOPT www.gams.com/latest/docs/S_DICOPT.html
- Filmint www.swmath.org/software/6197
- LaGO https://github.com/coin-or/LaGO
- LINDOGlobal www.lindo.com/index.php/products/lingo-and-optimization-modeling
- MindtPy https://pyomo.readthedocs.io/en/stable/contributed_packages/mindtpy.html
- MINOPT http://titan.princeton.edu/MINOPT/minopt.html
- Minotaur www.anl.gov/mcs/minotaur-toolkit-for-mixed-integer-nonlinear-optimization-problems
- MOSEK www.mosek.com/
- Pajarito https://github.com/JuliaOpt/Pajarito.jl
- SBB www.gams.com/latest/docs/S_SBB.html
- SCIP www.scipopt.org/
- SHOT https://github.com/coin-or/SHOT
- XPRESS www.fico.com/en/products/fico-xpress-optimization

GDP

- EMP www.gams.com/latest/docs/S_JAMS.html#EMP_DISJUNCTIVE_PROGRAMS
- LOGMIP www.logmip.ceride.gov.ar/
- Pyomo.GDP https://pyomo.readthedocs.io/en/stable/modeling_extensions/gdp.html

NEOS Server

https://neos-server.org/neos/

The NEOS Server is a free internet-based service for solving numerical optimization problems.

A.2.3 References on Optimization Software

Abhishek, K., Leyffer, S. and Linderoth, J. (2010) "FilMINT: An outer approximation-based solver for convex mixed-integer nonlinear programs", *INFORMS Journal on Computing*. doi: 10.1287/ijoc.1090.0373.

Aps, M. (2020) "MOSEK documentation".

Bandler, J. W., Charalambous, C. and Chen, J. H. K. (1976) "MINOPT – An optimization program based on recent minimax results (computer program descriptions)", *IEEE Transactions on Microwave Theory and Techniques*. doi: 10.1109/tmtt.1976.1128901.

Belotti, P. (2010) *COUENNE: A User's Manual*. Available at: https://projects.coin-or.org/Couenne/browser/trunk/Couenne/doc/couenne-user-manual.pdf?format=raw (Accessed: 20 May 2018).

Belotti, P., Berthold, T. and Neves, K. (2016) "Algorithms for discrete nonlinear optimization in FICO Xpress", in *Proceedings of the IEEE Sensor Array and Multichannel Signal Processing Workshop*. doi: 10.1109/SAM.2016.7569658.

Bernal, D. E. et al. (2018) "Mixed-integer nonlinear decomposition toolbox for Pyomo (MindtPy)", in *Computer Aided Chemical Engineering*, pp. 895–900. doi: 10.1016/B978-0-444-64241-7.50144-0.

Bernal, D. E. et al. (2020) "Improving the performance of DICOPT in convex MINLP problems using a feasibility pump", *Optimization Methods and Software*. doi: 10.1080/10556788.2019.1641498.

Bonami, P. et al. (2008) "An algorithmic framework for convex mixed integer nonlinear programs", *Discrete Optimization*, 5(2), 186–204. doi: 10.1016/j.disopt.2006.10.011.

Bussieck, M. R. and Drud, A. S. (2001) "SBB: A new solver for mixed integer nonlinear programming", in *Talk at OR2001*.

Coey, C., Lubin, M. and Vielma, J. P. (2020) "Outer approximation with conic certificates for mixed-integer convex problems", *Mathematical Programming Computation*. doi: 10.1007/s12532-020-00178-3.

Conn, A., Gould, N. I. M. and Toint, P. L. (1992) "LANCELOT: a Fortran package for large-scale nonlinear optimization (Release A)", En.Scientificcommons.Org.

Cook, W. et al. (2011) "An exact rational mixed-integer programming solver", in *Lecture Notes in Computer Science* (including subseries *Lecture Notes in Artificial Intelligence and Lecture Notes in Bioinformatics*). doi: 10.1007/978-3-642-20807-2_9.

Drud, A. S. (1994) "CONOPT – A large-scale GRG code", *ORSA Journal on Computing*. doi: 10.1287/ijoc.6.2.207.

Ferris, M. C. et al. (2009) "An extended mathematical programming framework", *Computers and Chemical Engineering*, 33(12), 1973–1982.

FICO (2020) "FICO®Xpress Optimization Suite". FICO San Jose, CA, USA.

Forrest, J. and Lougee-Heimer, R. (2005) "CBC user guide", in *Emerging Theory, Methods, and Applications*. doi: 10.1287/educ.1053.0020.

Gamrath, G. et al. (2020) The SCIP Optimization Suite 7.0.

Gill, P. E., Murray, W. and Saunders, M. A. (2005) "SNOPT: An SQP algorithm for large-scale constrained optimization", *SIAM*. doi: 10.1137/S1052623499350013.

Grossmann, I. E. et al. (2002) "GAMS/DICOPT: A Discrete Continuous Optimization Package".

Gurobi Optimization, L. L. C. (2020) "Gurobi Optimizer Reference Manual".

Hart, W. E. et al. (2017) *Pyomo – Optimization Modeling in Python*. Cham: Springer International Publishing (Springer Optimization and Its Applications). doi: 10.1007/978-3-319-58821-6.

IBM (2020) CPLEX Optimizer, IBM.

Kronqvist, J., Lundell, A. and Westerlund, T. (2016) "The extended supporting hyperplane algorithm for convex mixed-integer nonlinear programming", *Journal of Global Optimization*, **64**(2), 249–272. doi: 10.1007/s10898-015-0322-3.

Kronqvist, J. et al. (2019) "A review and comparison of solvers for convex MINLP", *Optimization and Engineering*. Springer, pp. 1–59. doi: 10.1007/s11081-018-9411-8.

Lin, Y. and Schrage, L. (2009) "The global solver in the LINDO API", *Optimization Methods and Software*. doi: 10.1080/10556780902753221.

Misener, R. and Floudas, C. A. (2014) "ANTIGONE: Algorithms for coNTinuous / Integer Global Optimization of Nonlinear Equations", *Journal of Global Optimization*, **59**(2–3), 503–526. doi: 10.1007/s10898-014-0166-2.

Mittelmann, H. D. (2020) "Benchmarking optimization software – a (Hi)Story", *SN Operations Research Forum*, **1**(1), 2. doi: 10.1007/s43069-020-0002-0.

Nowak, I., Alperin, H. and Vigerske, S. (2003) "LaGO – An object oriented library for solving MINLPs", in Bliek C., Jermann C. and Neumaier A. (eds.) *Global Optimization and Constraint Satisfaction. COCOS 2002. Lecture Notes in Computer Science*, vol. 2861. Berlin, Heidelberg: Springer. doi: 10.1007/978-3-540-39901-8_3.

Oki, E. (2012) "GLPK (GNU Linear Programming Kit)", in *Linear Programming and Algorithms for Communication Networks*. doi: 10.1201/b12733-4.

Pintér, J. D. (2007) "Nonlinear optimization with GAMS /LGO", *Journal of Global Optimization*. doi: 10.1007/s10898-006-9084-2.

Tawarmalani, M. and Sahinidis, N. V. (2002) *Convexification and Global Optimization in Continuous and Mixed-Integer Nonlinear Programming*. Springer. doi: 10.1007/978-1-4757-3532-1.

Vanderbei, R. J. (1999) "LOQO: An interior point code for quadratic programming", *Optimization Methods and Software*, **11**(1–4), 451–484. doi: 10.1080/10556789908805759.

Vecchietti, A. and Grossmann, I. E. (1999) "LOGMIP: A disjunctive 0-1 non-linear optimizer for process system models", *Computers and Chemical Engineering*, **23**, 555–565. doi: 10.1016/S0098-1354(98)00293-2.

Viswanathan, J. and Grossmann, I. E. (1990) "A combined penalty function and outer approximation method for MINLP optimization", *Computers and Chemical Engineering*, **14**, 769. doi.org/10.1016/0098-1354(90)87085-4.

Wächter, A. and Biegler, L. T. (2006) "On the implementation of an interior-point filter line-search algorithm for large-scale nonlinear programming", *Mathematical Programming*, **106**, 25–57. doi: 10.1007/s10107-004-0559-y.

Westerlund, T. and Pettersson, F. (1995) "An extended cutting plane method for solving convex MINLP problems", *Computers and Chemical Engineering*, **19** (Suppl. 1), 131–136. doi: 10.1016/0098-1354(95)87027-X.

A.2.4 Libraries Test Problems

Mittleman Collection of NLP Problems

http://plato.asu.edu/ftp/ampl-nlp.html

CUTE Test Problems

www.cuter.rl.ac.uk/problems.html
www.cuter.rl.ac.uk/Problems/mastsif.shtml

MINLPLib: A Library of Mixed-Integer and Continuous Nonlinear Programming Instances

www.minlplib.org/

MINLP (instances minlp.org)

www.minlp.org/library/instances.php

CMU-IBM Open Source MINLP Project

http://egon.cheme.cmu.edu/ibm/page.htm

SIPLIB: A Stochastic Integer Programming Test Problem Library

www2.isye.gatech.edu/~sahmed/siplib/

APPENDIX B

Optimization Models for Process Systems Engineering

B.1 Web-Based Software for Mixed-Integer Programming Applications in Process Systems Engineering

http://newton.cheme.cmu.edu/interfaces/

This weblink, developed at Carnegie Mellon, provides interfaces to problems in the areas of process synthesis, and planning and scheduling of process systems, through novel mathematical programming approaches, which rely on linear and nonlinear models with discrete and continuous variables. These include mixed-integer programming (MILP and MINLP), general disjunctive programming (GDP), and global-optimization multiperiod optimization. Both deterministic models as well as models with uncertainty are considered in the list below.

BatchMPC

Design of multiproduct batch plants with mixed-product campaigns. *(MILP)*

Voudouris, V. and Grossmann, I. E. (1992) "Mixed-integer linear programming models for the optimal design of batch processes", *Industrial & Engineering Chemistry Research*, **31**, 1315–1325.

BatchSPC

Design of multiproduct batch plants with single-product campaigns. *(MILP, MINLP)*

Kocis, G. R. and Grossmann, I. E. (1988) "Global optimization of nonconvex MINLP problems in process synthesis", *Industrial & Engineering Chemistry Research*, **27**, 1407.

Voudouris, V. and Grossmann, I. E. (1993) "Optimal synthesis of multiproduct batch plants with cyclic scheduling and inventory considerations", *Industrial & Engineering Chemistry Research*, **32**, 1962–1980.

CENTRALIZED/DECENTRALIZED

Generalized disjunctive program model for the optimal location of centralized and decentralized manufacture facilities. *(GDP, MINLP)*

Lara, C. L. and Grossmann, I. E. (2016) "Global optimization for a continuous location-allocation model for centralized and distributed manufacturing", *Computer Aided Chemical Engineering*, **38**, 1009–1014.

Lara, C. L., Trespalacios, F. and Grossmann, I. E. (2018) "Global optimization algorithm for capacitated multi-facility continuous location-allocation problems", *Journal of Global Optimization*, **71**, 871–889.

CHP

Combined heat and power: optimization of heat and power plants under time-sensitive electricity prices. *(MILP)*

Mitra, S., Sun, L. and Grossmann, I. E. (2013) "Optimal scheduling of industrial combined heat and power plants under time-sensitive electricity prices", *Energy*, **54**, 194–211.

CRUDEOIL

Inventory management of a refinery that imports several types of crude oil delivered by different vessels.

CRUDEOIL DISCRETE: *(MILP)*
Lee, H., Pinto, J. M., Grossmann, I. E. and Park, S. (1996) "MILP model for refinery short term scheduling of crude oil unloading with inventory management", *Industrial & Engineering Chemistry Research*, **35**, 1630–1641.

CRUDEOIL CONTINUOUS: *(MINLP)*
Mouret, S., Grossmann, I. E. and Pestiaux, P. (2009) "A novel priority-slot based continuous-time formulation for crude-oil scheduling problems", *Industrial & Engineering Chemistry Research*, **48** (18), 8515–8528.

CYCLE

Zero-wait scheduling of multi-product batch plants. *(LP, MINLP)*

Birewar, D. and Grossmann, I. E. (1989) "Efficient optimization algorithms for zero-wait scheduling of multiproduct bath plants", *Industrial & Engineering Chemistry Research*, **28**, 1333.

DECAY

Scheduling multiple feeds on parallel units, where the performance of each unit decreases with time. *(MINLP)*

Jain, V. and Grossmann, I. E. (1998) "Cyclic scheduling and maintenance of parallel process units with decaying performance", *AIChE Journal*, **44**, 1623–1636.

DISIM

Distribution center design with stochastic inventory management. *(MINLP)*

You, F. and Grossmann, I. E. (2008) "Mixed-integer nonlinear programming models and algorithms for large-scale supply chain design with stochastic inventory management", *Industrial & Engineering Chemistry Research*, **47**, 7802–7817.

EXTRACTOR

Optimal design of multicomponent liquid–liquid extraction processes using multistage countercurrent extractor systems. *(GDP)*

Reyes-Labarta, J. A. and Grossmann, I. E. (2001) "Disjunctive programming models for the optimal design of complex liquid–liquid multistage extractors", *AIChE Journal*, **47**, 2243–2252.

FLEXNET

Flexibility index evaluation for heat-exchanger networks. *(MILP, MINLP)*

Escobar, M. and Grossmann, I. E. (2011) "SynFlex: A computational framework for synthesis of flexible heat exchanger networks", *Computer Aided Chemical Engineering*, **29**, 1924–1928.

GDP-DISTILL

Synthesizing a single distillation column. *(GDP)*

Barttfeld, M., Aguirre, P. A. and Grossmann, I. E. (2003) "Alternative representations and formulations for the economic optimization of multicomponent distillation columns", *Computers and Chemical Engineering*, **27**, 363–383.

GREENPLAN

Bicriterion optimization of industrial chemical complexes. *(MILP)*

Grossmann, I. E., Drabbant, R. and Jain, R. K. (1982) "Incorporating toxicology in the synthesis of industrial chemical complexes", *Chemical Engineering Communications*, **17**(1–6), 151–170.

LOGMIP

Logic mixed-integer programming solver. *(GDP)*

MINLP Cyberinfrastructure

Collaborative site with library of MINLP problems and their formulation. *(MINLP, MILP)*

MULTIPERIOD BLEND

Decomposition method for multiperiod blending problem. *(MINLP)*

Trespalacios, F. and Grossmann, I. E. (2016) "An MILP-MINLP decomposition method for the global optimization of a source based model of the multiperiod blending problem", *Computers and Chemical Engineering*, **87**, 13–35.

MULTISTAGE

Optimization of cyclic schedules of multiproduct continuous plants. *(MINLP)*

Pinto, J. M. and Grossmann, I. E. (1994) "Optimal scheduling of multistage multiproduct continuous plants", *Computers and Chemical Engineering*, **9**, 797–816.

NETCHAIN

Optimization of the supply chain in continuous flexible process networks. *(MILP)*

Bok, J. K., Grossmann, I. E. and Park, S. (2000) "Supply chain optimization in continuous flexible process networks", *Industrial & Engineering Chemistry Research*, **39**(5), 1279–1290.

OILGASPLAN

Planning of offshore oilfield development. *(MINLP)*

Gupta, V. and Grossmann, I. E. (2012) "An efficient multiperiod MINLP model for optimal planning of offshore oil and gas infrastructure", *Industrial & Engineering Chemistry Research*, **51**(19), 6823–6840.

PowerSCHEDULE

Simultaneous optimization of production scheduling and electricity procurement for continuous power-intensive processes. *(MILP)*

Zhang, Q., Cremer, J. L., Grossmann, I. E., Sundaramoorthy, A. and Pinto, J. M. (2016) "Risk-based integrated production scheduling and electricity procurement for continuous power-intensive processes", *Computers and Chemical Engineering*, **86**, 90–105.

PRODEV

Scheduling of testing tasks in the new product development. *(MILP)*

Maravelias, C. T. and Grossmann, I. E. (2001) "Simultaneous planning for new product development and batch manufacturing facilities", *Industrial & Engineering Chemistry Research*, **40**, 6147–6164.

Schmidt, C. W. and Grossmann, I. E. (1996) "Optimization models for the scheduling of testing tasks in new product development", *Industrial & Engineering Chemistry Research*, **35**, 3498–3510.

REL OPT

Optimal Design of Reliable Chemical Plants *(MINLP)*

Ye, Y., I. E. Grossmann and J. M. Pinto (2018) "Mixed-integer nonlinear programming models for optimal design of reliable chemical plants", *Computers and Chemical Engineering*, **116**, 3–16.

Resilient Supply Chain

Supply chain with risk of facility disruptions. *(MILP)*

Garcia-Herreros, P., Wassick, J. M. and Grossmann, I. E. (2014) "Design of resilient supply chains with risk of facility disruptions", *Industrial & Engineering Chemistry Research*, **54**, 17240–17251.

SIMPLAN

Simultaneous planning and scheduling of a single-stage multiproduct. *(MILP)*

Erdirik-Dogan, M. and Grossmann, I. E. (2008) "Simultaneous planning and scheduling of single-stage multi-product continuous plants with parallel lines", *Computers and Chemical Engineering*, **32** (11), 2664–2683.

SIMPRONET

Stochastic inventory management for process networks. *(MINLP)*

You, F. and Grossmann, I. E. (2011) "Stochastic inventory management for tactical process planning under uncertainties: MINLP models and algorithms", *AIChE Journal*, **57**, 1250–1277.

STEAM

Optimal synthesis and operation of utility plants. *(MINLP)*

Bruno, J. C. and Grossmann, I. E. (1998) "A rigorous MINLP model for the optimal synthesis and operation of utility plants", *Chemical Research and Design,* **76**(3), 246–258.

STN

State Task Network, continuous and discrete models. *(MILP)*

For discrete-time model

Kondili, E., Pantelides, C. C. and Sargent, R. W. H. (1993) "A general algorithm for short-term scheduling of batch operations – I. MILP formulation", *Computers and Chemical Engineering*, **2**, 211–227.

Shah, N., Pantelides, C. C. and Sargent, R. W. H. (1993) "A general algorithm for short-term scheduling of batch operations – II. Computational issues", *Computers and Chemical Engineering*, **2**, 229–244.

For continuous-time model

Maravelias, C. T. and Grossmann, I. E. (2003) "New continuous-time state task network formulation for the scheduling of multipurpose batch plant", *Industrial & Engineering Chemistry Research*, **42** (13), 3056–3074.

SYNHEAT

Program for optimizing heat-exchanger networks. *(MINLP)*

Ponce-Ortega, J. M., Jimenez-Gutierrez, A. and Grossmann, I. E. (2007) "Optimal synthesis of heat exchanger networks involving isothermal process streams", *Computers and Chemical Engineering*, **32**(8), 1918–1942.

Yee, T. F. and Grossmann, I. E. (1990) "Simultaneous optimization model for heat integration – II. Heat exchanger network synthesis", *Computers and Chemical Engineering*, **14** (10), 1165–1184.

THERMAL-DIST

Optimal separation sequences based on thermally coupled distillation.

Caballero, J. and Grossmann, I. E. (2004) "Design of distillation sequences: from conventional to fully thermally coupled distillation systems", *Computers and Chemical Engineering*, **28**(11), 2307–2329.

WATER

Design of integrated and distributed wastewater networks. *(Global NLP)*

WATERNET

Karuppiah, R. and Grossmann, I. E. (2006) "Global optimization for the synthesis of integrated water systems in chemical processes", *Computers and Chemical Engineering*, **30**(4), 650–673.

WATERTREAT

Galan, B. and Grossmann, I. E. (1998) "Optimal design of distributed wastewater treatment networks", *Industrial & Engineering Chemistry Research*, **37**(10), 4036–4048.

B.2   CMU-IBM Cyber-Infrastructure for MINLP Collaborative Site

www.minlp.org/

This collaborative site, developed at Carnegie Mellon, contains optimization models in different areas contributed by researchers in the optimization field. The site contains linear and nonlinear models with one or several alternative model formulations involving discrete and continuous variables through mixed-integer nonlinear programming (MINLP), mixed-integer linear programming (MILP), or generalized disjunctive programming (GDP).

MINLP Library

	Problem Title	Primary Author	Model Types
1	AC Network-Constrained Unit Commitment	Anjos, Miguel	MINLP
2	Global Optimization Algorithm for Capacitated Multi-facility Continuous Location-Allocation Problems	Lara, Cristiana	MINLP
3	Mixed-integer nonlinear programming models for optimal design of reliable chemical plants	Ye, Yixin	MINLP
4	Crude Oil Pooling Problem	Castro, Pedro	MINLP, MILP
5	Industrial gas network operation	Puranik, Yash	MINLP
6	Optimal Procurement Contract Selection with Price Optimization under Uncertainty for Process Networks	Calfa, Bruno	MINLP
7	Large-Scale MILP Transshipment Models for Heat Exchanger Network Synthesis	Chen, Yang	MILP
8	Mixed-Integer Linear Fractional Programming	You, Fengqi	MINLP, LP, MILP
9	Hydro Energy System Scheduling	Castro, Pedro	MILP, MIQCP
10	Chance-constrained problems with quadratic constraints	Margot, Francois	MINLP
11	Sustainable Integration of Trigeneration Systems with Heat Exchanger Networks	Ponce-Ortega, Jose Maria	MINLP
12	Cyclic Scheduling with Heat Sharing	Castro, Pedro	MILP
13	Multiperiod Blend Scheduling Problem	Trespalacios, Francisco	MILP, MIQCP
14	On the solution of 0-1 Quadratic Programming Problems with cardinality constraints	Lima, Ricardo	MINLP, MILP, MIQP
15	The optimal design of a three-echelon supply chain with inventories under uncertainty	Nyberg, Axel	MINLP, MILP
16	A Novel Bi-Level Optimization Method for the Identification of Critical Points in Flow Sheet Synthesis under Uncertainty	Novak Pintaric, Zorka	NLP
17	Continuous-time Representations in Scheduling	Castro, Pedro	MILP
18	Optimal Design of Water Distribution Networks	D'Ambrosio, Claudia	MINLP, MILP
19	MINLP and MPCC formulations for the cascading tanks problem	Gopalakrishnan, Ajit	MINLP, NLP
20	Optimal solvent design for reactions using computer-aided molecular design	Struebing, Heiko	MILP

(cont.)

	Problem Title	Primary Author	Model Types
21	The kissing number problem	Belotti, Pietro	QP
22	2-D Orthogonal Strip Packing	Castro, Pedro	MILP
23	Mixed-Integer Linear Programming Models for Scheduling in Process Networks	Sundaramoorthy, Arul	MILP
24	MINLP approach for the TSPN (Traveling Salesman Problem with Neighborhoods)	Gentilini, Iacopo	MINLP
25	Generalized Pooling Problem	Misener, Ruth	MIQCP
26	Crude-oil Operations Scheduling	Mouret, Sylvain	MINLP
27	Inventory-Production-Distribution Problems with Direct Shipments	Margot, Francois	MILP
28	Optimal Scheduling of Multistage Batch Plants	Castro, Pedro	MILP
29	Integrated Process Water Networks Design Problem	Ahmetović, Elvis	MINLP
30	Cutting Stock Optimization Problem for the Production of Carton Board Boxes	Vecchietti, Aldo	MINLP, MILP
31	Mixed-Integer Nonlinear Programming Models for Optimal Simultaneous Synthesis of Heat Exchangers Network	Escobar, Marcelo	MINLP
32	Optimal Scheduling of Refined Products Pipelines and Terminal Operations	Cafaro, Diego	MILP
33	MPEC strategies for optimization of pipeline operations	Gopalakrishnan, Ajit	NLP
34	Extended Pooling Problem with the Summer Time Complex Emissions Constraints	Misener, Ruth	MINLP
35	Optimization of metabolic networks in biotechnology	Guillen, Gonzalo	MINLP
36	Close-Enough Vehicle Routing Problem	Mennell, William	MINLP
37	Simultaneous Cyclic Scheduling and Control of a Multiproduct CSTR	Flores-Tlacuahuac, Antonio	MINLP
38	Optimal Separation Sequences Based on Distillation: From Conventional to Fully Thermally Coupled Systems	Caballero, Jose	MINLP
39	The Close-Enough Traveling Salesman Problem	Mennell, William	MINLP
40	The Delay Constrained Routing Problem (DCRP)	Hijazi, Hassan	MINLP
41	A Deterministic Security Constrained Unit Commitment Model	Zondervan, Edwin	MINLP, MIQP

(cont.)

	Problem Title	Primary Author	Model Types
42	Stochastic Portfolio Optimization with Round Lot Trading Constraints	Lejeune, Miguel	MINLP
43	Stabilizing controller design and the Belgian chocolate problem	Chang, YoungJung	MINLP
44	Design of Telecommunication Networks with Shared Protection	Belotti, Pietro	MINLP, MILP
45	Mixed-Integer Nonlinear Programming Models for the Optimal Design of Multi-product Batch Plant	You, Fengqi	MINLP
46	Solving Mixed-Integer Linear Fractional Programming Problems with Dinkelbach's Algorithm and MINLP Methods	You, Fengqi	MINLP
47	Global multi-objective optimization of a nonconvex MINLP problem and its application on polygeneration energy systems design	Liu, Pei	MINLP
48	Periodic Scheduling of Continuous Multiproduct Plants	Castro, Pedro	MINLP
49	Optimization model for density modification based on single-crystal X-ray diffraction data	Sahinidis, Nick	MINLP, MILP
50	Mixed-Integer Nonlinear Programming Models and Algorithms for Supply Chain Design with Stochastic Inventory Management	You, Fengqi	MINLP
51	MINLP & MPCC Strategies for Optimization of a Class of Hybrid Dynamic Systems	Baumrucker, Brian	MINLP, NLP, MIQP
52	Water Treatment Network Design	Ruiz, Juan	MINLP

GDP Library

	Problem Title	Primary Author	Model type
1	Separation of fiber and stickies	Trespalacios, Francisco	Nonlinear
2	Design of multi-product batch plant	Trespalacios, Francisco	Nonlinear
3	Medium-Term Purchasing Contracts	Vecchietti, Aldo	Linear
4	Process Synthesis Problem	Ruiz, Juan	Nonlinear
5	Jobshop Scheduling	Vecchietti, Aldo	Linear
6	Strip Packing	Vecchietti, Aldo	Linear
7	2-D Constrained Layout	Ruiz, Juan	Nonlinear

References

Adjiman, C. S., Dallwig, S., Floudas, C. A. and Neumaier, A. (1998) "A global optimization method, αBB, for general twice-differentiable constrained NLPs – I. Theoretical advances", *Computers & Chemical Engineering*, **22**(9), 1137–1158. doi: 10.1016/S0098-1354(98)00027-1.

Alonso-Ayuso, A., Escudero, L. F, Garín, A., Ortuño, M. T. and Pérez, G. (2003) "An approach for strategic supply chain planning under uncertainty based on stochastic 0-1 programming", *Journal of Global Optimization*, **26**, 97–124. doi.org/10.1023/A:1023071216923

Applegate, D. and Cook, W. (1991) "A computational study of the job-shop scheduling problem", *ORSA Journal on Computing*, **3**(2), 149–156. doi: 10.1287/ijoc.3.2.149.

Audet, C., Savard, G. and Zghal, W. (2008) "Multiobjective optimization through a series of single-objective formulations", *SIAM Journal on Optimization*, **19**(1), 188–210. doi: 10.1137/060677513.

Balas, E. (1979) "Disjunctive programming", *Annals of Discrete Mathematics*, **5**, 3–51. doi: 10.1016/S0167-5060(08)70342-X.

Balas, E. (1985) "Disjunctive programming and a hierarchy of relaxations for discrete optimization problems", *SIAM Journal on Algebraic Discrete Methods*, **6**(3), 466–486. doi: 10.1137/0606047.

Balas, E. (2010) "Disjunctive programming", in *50 Years of Integer Programming 1958–2008*. Berlin, Heidelberg: Springer, pp. 283–340. doi: 10.1007/978-3-540-68279-0_10.

Balas, E. (2018) *Disjunctive Programming*. Springer Nature Switzerland.

Balas, E., Ceria, S. and Cornuéjols, G. (1993) "A lift-and-project cutting plane algorithm for mixed 0–1 programs", *Mathematical Programming*, **58**(1–3), 295–324. doi: 10.1007/BF01581273.

Bazaraa, M. S., Sherali, H. D. and Shetty, C. M. (2006) *Nonlinear Programming: Theory and Algorithms*. Hoboken, NJ: John Wiley & Sons, Inc. doi: 10.1002/0471787779.

Beale, E. M. L. (1979) "Branch and bound methods for mathematical programming systems", *Annals of Discrete Mathematics*, **5**, 201–219. doi: 10.1016/S0167-5060(08)70351-0.

Beale, E. M. L. and Forrest, J. J. H. (1976) "Global optimization using special ordered sets", *Mathematical Programming*, **10**(1), 52–69. doi: 10.1007/BF01580653.

Ben-Tal, A. and Nemirovski, A. (1999) "Robust solutions of uncertain linear programs", *Operations Research Letters*, **25**(1), 1–13. doi: 10.1016/S0167-6377(99)00016-4.

Ben-Tal, A., Goryashko, A., Guslitzer, E. and Nemirovski, A. (2004) "Adjustable robust solutions of uncertain linear programs", *Mathematical Programming*, **99**(2), 351–376. doi: 10.1007/s10107-003-0454-y.

Benders, J. F. (1962) "Partitioning procedures for solving mixed-variables programming problems", *Numerische Mathematik*, **4**(1), 238–252. doi: 10.1007/BF01386316.

Bertsekas, D. (1998) *Network Optimization: Continuous and Discrete Models*. Belmont, Massachusetts: Athena Scientific.

Bertsimas, D., Brown, D. B. and Caramanis, C. (2011) "Theory and applications of robust optimization", *SIAM Review*, **53**(3), 464–501. doi: 10.1137/080734510.

Biegler, L. T. (2010) *Nonlinear Programming*. Society for Industrial and Applied Mathematics. doi: 10.1137/1.9780898719383. x

Biegler, L. T. and Grossmann, I. E. (2004) "Retrospective on optimization", *Computers*

& *Chemical Engineering*, **28**(8), 1169–1192. doi: 10.1016/j.compchemeng.2003.11.003.

Biegler, L. T., Grossmann, I. E. and Westerberg, A. W. (1997) *Systematic Methods of Chemical Process Design*, New Jersey.

Birge, J. R. and Louveaux, F. (2011) *Introduction to Stochastic Programming* (Springer Series in Operations Research and Financial Engineering). New York: Springer. doi: 10.1007/978-1-4614-0237-4.

Bixby, R. and Rothberg, E. (2007) "Progress in computational mixed integer programming – A look back from the other side of the tipping point", *Annals of Operations Research*, **149**(1), 37–41. doi: 10.1007/s10479-006-0091-y.

Bonami, P., Kilinç, M. and Linderoth, J. (2012) *Mixed Integer Nonlinear Programming* (The IMA Volumes in Mathematics and its Applications), Lee, J. and Leyffer, S. (eds.), New York: Springer. doi: 10.1007/978-1-4614-1927-3.

Bottou, L., Curtis, F. E. and Nocedal, J. (2018) "Optimization methods for large-scale machine learning", *SIAM Review*, **60**(2), 223–311. doi.org/10.1137/16M1080173.

Broyden, C. G. (1965) "A class of methods for solving nonlinear simultaneous equations", *Mathematics of Computation*, **19**(92), 577. doi: 10.2307/2003941.

Byrd, R. H., Nocedal, J. and Waltz, R. A. (2006) "KNITRO: An integrated package for nonlinear optimization", in Di Pillo, G. and Roma, M. (eds.) *Large-Scale Nonlinear Optimization. Nonconvex Optimization and Its Applications*, vol. 83. Boston, MA: Springer.

Carøe, C. C. and Schultz, R. (1999) "Dual decomposition in stochastic integer programming", *Operations Research Letters*, **24**(1–2), 37–45. doi: 10.1016/S0167-6377(98)00050-9.

Castro, P. M. and Grossmann, I. E. (2012) "Generalized disjunctive programming as a systematic modeling framework to derive scheduling formulations", *Industrial & Engineering Chemistry Research*, **51**, 5781–5792. doi: 10.1021/ie2030486.

Ceria, S. and Soares, J. (1999) "Convex programming for disjunctive convex optimization",

Mathematical Programming, **86**(3), 595–614. doi: 10.1007/s101070050106.

Clocksin, W. F. and Mellish, C. S. (1994) *Programming in Prolog*. Berlin, Heidelberg: Springer-Verlag. doi: 10.1007/978-3-642-97596-7.

Conforti, M., Cornuéjols, G. and Zambelli, G. (2014) "Integer programming models", in *Integer Programming*. Springer, pp. 45–84.

Cook, W. J. (2012) *In Pursuit of the Traveling Salesman: Mathematics at the Limits of Computation*. Princeton University Press.

Cornuéjols, G., Nemhauser, G. A. and Wolsey, L. A. (1990) "The uncapacitated facility location problem", in Mirchandani, P. B. and Francis, R. L. (eds.) *Discrete Location Theory*, New York: Wiley.

Crowder, H., Johnson, E. L. and Padberg, M. (1983) "Solving large-scale zero-one linear programming problems", *Operations Research*, **31**(5), 803–834. doi: 10.1287/opre.31.5.803.

Dakin, R. J. (1965) "A tree-search algorithm for mixed integer programming problems", *The Computer Journal*, **8**(3), 250–255. doi: 10.1093/comjnl/8.3.250.

Dantzig, G. (1963) *Linear Programming and Extensions*. RAND Corporation. doi: 10.7249/R366.

Dantzig, G. B. (1949) "Programming in a linear structure", *Econometrica*, **17**, 73–74.

Drud, A. S. (1994) "CONOPT – A large-scale GRG code", *ORSA Journal on Computing*, **6**(2), 207–216. doi: 10.1287/ijoc.6.2.207.

Duffin, R. J., Peterson, E. L. and Zener, C. (1967) *Geometric Programming – Theory and Applications*. New York: John Wiley and Sons.

Duran, M. A. and Grossmann, I. E. (1986) "An outer-approximation algorithm for a class of mixed-integer nonlinear programs", *Mathematical Programming*, **36**(3), 307–339. doi: 10.1007/BF02592064.

Eiben, A. E. and Smith, J. E. (2003) *Introduction to Evolutionary Computing*, Springer.

Fiacco, A. V. and McCormick, G. P. (1968) *Nonlinear Programming: Sequential Unconstrained Minimization Techniques*. New York: John Wiley and Sons.

Fisher, M. L. (1981) "The Lagrangian relaxation method for solving integer programming problems", *Management Science*, **27**(1), 1–18. doi: 10.1287/mnsc.27.1.1.

Floudas, C. A. (1995) *Nonlinear and Mixed-Integer Optimization: Fundamentals and Applications*. Oxford University Press.

Floudas, C. A. and Gounaris, C. E. (2009) "A review of recent advances in global optimization", *Journal of Global Optimization*, **45**(1), 3–38. doi: 10.1007/s10898-008-9332-8.

Frangioni, A. (2005) "About Lagrangian methods in integer optimization", *Annals of Operations Research*, **139**(1), 163–193. doi: 10.1007/s10479-005-3447-9.

Furman, K. C., Sawaya, N. W. and Grossmann, I. E. (2020) "A computationally useful algebraic representation of nonlinear disjunctive convex sets using the perspective function", *Computational Optimization and Applications*, **76**(2), 589–614. doi: 10.1007/s10589-020-00176-0.

Geoffrion, A. M. (1972) "Generalized Benders decomposition", *Journal of Optimization Theory and Applications*, **10**(4), 237–260. doi: 10.1007/BF00934810.

Geoffrion, A. M. (1974) "Lagrangean relaxation for integer programming", in Balinski, M. L. (ed.) *Approaches to Integer Programming. Mathematical Programming Studies, vol. 2.* Berlin, Heidelberg: Springer, pp. 82–114. doi: 10.1007/BFb0120690.

El Ghaoui, L. (1997) "Robust optimization", *IFAC Proceedings Volumes*, **30**(16), 115–117. doi: 10.1016/S1474-6670(17)42591-2.

Gomory, R. E. (1958) "Outline of an algorithm for integer solutions to linear programs", *Bulletin of the American Mathematical Society*, **64**(5), 275–279. doi: 10.1090/S0002-9904-1958-10224-4.

Gordan, P. (1873) "Ueber die Auflösung linearer Gleichungen mit reellen Coefficienten", *Mathematische Annalen*, **6**(1), 23–28. doi: 10.1007/BF01442864.

Goyal, V. and Ierapetritou, M. G. (2002) "Determination of operability limits using simplicial approximation", *AIChE Journal*, **48**(12), 2902–2909. doi: 10.1002/aic.690481217.

Griewank, A. (2000) *Evaluating Derivatives: Principles and Techniques of Algorithmic Differentiation*. Philadelphia: SIAM.

Griewank, A. and Walther, A. (2008) *Evaluating Derivatives*. Society for Industrial and Applied Mathematics. doi: 10.1137/1.9780898717761.

Grossmann, I. E. (2002) "Review of nonlinear mixed-integer and disjunctive programming techniques", *Optimization and Engineering*, **3**(3), 227–252. doi: 10.1023/A:1021039126272.

Grossmann, I. E. and Biegler, L. T. (2004) "Part II. Future perspective on optimization", *Computers & Chemical Engineering*, **28**(8), 1193–1218. doi: 10.1016/j.compchemeng.2003.11.006.

Grossmann, I. E., Calfa, B. A. and Garcia-Herreros, P. (2014) "Evolution of concepts and models for quantifying resiliency and flexibility of chemical processes", *Computers & Chemical Engineering*, **70**, 22–34. doi: 10.1016/j.compchemeng.2013.12.013.

Grossmann, I. E. and Floudas, C. A. (1987) "Active constraint strategy for flexibility analysis in chemical processes", *Computers & Chemical Engineering*, **11**(6), 675–693. doi: 10.1016/0098-1354(87)87011-4.

Grossmann, I. E. and Trespalacios, F. (2013) "Systematic modeling of discrete-continuous optimization models through generalized disjunctive programming", *AIChE Journal*, **59**, 3276–3295. doi: 10.1002/aic.14088.

Guignard, M. (2003) "Lagrangean relaxation", *TOP*, **11**, 51–228. doi: 10.1007/BF02579036.

Guignard, M. and Kim, S. (1987) "Lagrangean decomposition: A model yielding stronger Lagrangean bounds", *Mathematical Programming*, **39**(2), 215–228. doi: 10.1007/BF02592954.

Gupta, O. K. and Ravindran, A. (1985) "Branch and bound experiments in convex nonlinear integer programming", *Management Science*, **31**(12), 1533–1546. doi: 10.1287/mnsc.31.12.1533.

Haar, A. (1918) "Die Minkowskische Geometrie und die Annäherung an stetige Funktionen", *Mathematische Annalen*, **78**, 249–311.

Halemane, K. P. and Grossmann, I. E. (1983) "Optimal process design under uncertainty", *AIChE Journal*, **29**(3), 425–433. doi: 10.1002/aic.690290312.

Han, S.-P. (1976) "Superlinearly convergent variable metric algorithms for general nonlinear programming problems", *Mathematical Programming*, **11**(1), 263–282. doi: 10.1007/BF01580395.

Van Hentenryck, P. (1989) *Constraint Satisfaction in Logic Programming*. MIT Press.

Hettich, R. and Kortanek, K. O. (1993) "Semi-infinite programming: Theory, methods, and applications", *SIAM Review*, **35**(3), 380–429. doi: 10.1137/1035089.

Hiriart-Urruty, J.-B. and Lemaréchal, C. (2001) *Fundamentals of Convex Analysis*. Berlin, Heidelberg: Springer.

Hooker, J. N. (2002) "Logic, optimization, and constraint programming", *INFORMS Journal on Computing*, **14**(4), 295–321. doi: 10.1287/ijoc.14.4.295.2828.

Hooker, J. N. and van Hoeve, W.-J. (2018) "Constraint programming and operations research", *Constraints*, **23**, 172–195. doi: 10.1007/s10601-017-9280-3.

Horst, R. and Tuy, H. (1996) *Global Optimization: Deterministic Approaches*, Springer-Verlag.

John, F. (1948) "Extremum problems with inequalities as side conditions", in Friedrichs, K. O., Neugebauer, E. and Stoker, J. J. (eds.) *Studies and Essays, Courant Anniversary Volume*. New York: Wiley (Interscience), pp. 187–204.

Kantorovich, L. V. (1948) "On Newton's method for functional equations", *Doklady Akademii Nauk* (in Russian), **59**, 1237–1240.

Karmarkar, N. (1984) "A new polynomial-time algorithm for linear programming", *Combinatorica*, **4**(4), 373–395. doi: 10.1007/BF02579150.

Karush, W. (1939) *Minima of functions of several variables with inequalities as side constraints* (M.Sc. thesis). Department of Mathematics, University of Chicago, Chicago, Illinois.

Kesavan, P., Allgor, R. J., Gatzke, E. P. and Barton, P. I. (2004) "Outer approximation algorithms for separable nonconvex mixed-integer nonlinear programs", *Mathematical Programming*, **100**(3), 517–535. doi: 10.1007/s10107-004-0503-1.

Kocis, G. R. and Grossmann, I. E. (1987) "Relaxation strategy for the structural optimization of process flowsheets", *Industrial & Engineering Chemistry Research*, **26**, 1869–1880.

Kronqvist, J., Lundell, A. and Westerlund, T. (2016) "The extended supporting hyperplane algorithm for convex mixed-integer nonlinear programming", *Journal of Global Optimization*, **64**(2), 249–272. doi: 10.1007/s10898-015-0322-3.

Kronqvist, J., Bernal, D. E., Lundell, A. and Grossmann, I. E. (2019) "A review and comparison of solvers for convex MINLP", *Optimization and Engineering*, **20**(2), 397–455. doi: 10.1007/s11081-018-9411-8.

Kuhn, H. W. and Tucker, A. W. (1951) "Nonlinear programming", in *Proceedings of the Second Berkeley Symposium on Mathematical Statistics and Probability*. Berkeley: University of California Press, pp. 481–492.

Land, A. H. and Doig, A. G. (1960) "An automatic method of solving discrete programming problems", *Econometrica*, **28**(3), 497. doi: 10.2307/1910129.

Laporte, G. (1992) "The traveling salesman problem: An overview of exact and approximate algorithms", *European Journal of Operational Research*, **59**(2), 231–247. doi: 10.1016/0377-2217(92)90138-Y.

Lasdon, L. S., Waren, A. D., Jain, A. and Ratner, M. (1978) "Design and testing of a generalized reduced gradient code for nonlinear programming", *ACM Transactions on Mathematical Software*, **4**, 34–50. doi: 10.1145/355769.355773.

Lee, J. and Leyffer, S. (eds.) (2012) *Mixed Integer Nonlinear Programming* (The IMA Volumes in Mathematics and Its Applications). New York: Springer. doi: 10.1007/ 978-1-4614-1927-3.

Lee, S. and Grossmann, I. E. (2000) "New algorithms for nonlinear generalized disjunctive programming", *Computers & Chemical Engineering*, **24**(9–10), 2125–2141. doi: 10.1016/S0098-1354(00)00581-0.

Lin, X., Janak, S. L. and Floudas, C. A. (2004) "A new robust optimization approach for s cheduling under uncertainty: I. Bounded uncertainty", *Computers and Chemical Engineering* 28 1069–1085. doi:10.1016/j.compchemeng.2003.09.020

Lovász, L. and Schrijver, A. (1991) "Cones of matrices and set-functions and 0–1 optimization", *SIAM Journal on Optimization*, **1**(2), 166–190. doi: 10.1137/0801013.

Marsten, R., Subramanian, R., Saltzman, M. J. et al. (1990) "Interior point methods for linear programming: Just call Newton, Lagrange, and Fiacco and McCormick!", *Interfaces*, **20**(4), 105–116.

McCormick, G. P. (1976) "Computability of global solutions to factorable nonconvex programs: Part I - Convex underestimating problems", *Mathematical Programming*, **10**(1), 147–175.

Minoux, M. (1986) *Mathematical Programming: Theory and Algorithms*. John Wiley & Sons.

Misener, R. and Floudas, C. A. (2013) "GloMIQO: Global mixed-integer quadratic optimizer", *Journal of Global Optimization*, **57**(1), 3–50. doi: 10.1007/s10898-012-9874-7.

Morrison, D. R., Jacobson, S. H., Sauppe, J.J. and Sewell, E. W. (2016) "Branch-and-bound algorithms: A survey of recent advances in searching, branching, and pruning", *Discrete Optimization*, **19**, 79–102. doi: 10.1016/j.disopt.2016.01.005.

Murtagh, B. A. and Saunders, M. A. (1978) "Large-scale linearly constrained optimization", *Mathematical Programming*, **14**(1), 41–72. doi: 10.1007/BF01588950.

Murtagh, B. A. and Saunders, M. A. (1982) "A projected Lagrangian algorithm and its implementation for sparse nonlinear constraints", in Buckley, A. G. and Goffin, J.-L. (eds.) *Algorithms for Constrained Minimization of Smooth Nonlinear Functions*, Springer, pp. 84–117. doi: 10.1007/BFb0120949.

Murtagh, B. A. and Saunders, M. A. (2003) *MINOS 5.5 User's Guide*, Stanford University Systems Optimization Laboratory.

Nemhauser, G. and Wolsey, L. (1988) *Integer and Combinatorial Optimization*. Hoboken, NJ: John Wiley & Sons, Inc. doi: 10.1002/9781118627372.

Nocedal, J. (1980) "Updating quasi-Newton matrices with limited storage", *Mathematics of Computation*, **35**, 773–782. doi: 10.2307/2006193.

Nocedal, J. and Wright, S. (2006) *Numerical Optimization*. Springer Science & Business Media.

Ortega, J. M. and Rheinbolt, W. C. (1970) *Iterative Solution of Non-linear Equations in Several Variables*. New York: Academic Press.

Pistikopoulos, E. N. and Grossmann, I. E. (1988) "Optimal retrofit design for improving process flexibility in linear systems", *Computers & Chemical Engineering*, **12**(7), 719–731. doi: 10.1016/0098-1354(88)80010-3.

Pistikopoulos, E. N., Georgiadis, M. C. and Dua, V. (2007) *Multi-Parametric Programming: Volume 1: Theory, Algorithms, and Applications*, Wiley VCH-Verlag & Co.

Powell, M. J. D. (1978) "A fast algorithm for nonlinearly constrained optimization calculations", in Watson, G. A. (ed.) *Numerical Analysis, Proceedings of the Biennial Conference Held at Dundee, June 28–July 1, 1977*, Springer Lecture Notes in Mathematics, volume 630, pp. 144–157. doi: 10.1007/BFb0067703.

Prékopa, A. (1970) "On probabilistic constrained programming", in *Proceedings of the Princeton Symposium on Mathematical Programming*. Princeton: Princeton University Press, pp. 113–138. doi: 10.1515/9781400869930-009.

Quesada, I. and Grossmann, I. E. (1992) "An LP/NLP based branch and bound algorithm for convex MINLP optimization problems", *Computers & Chemical Engineering*, **16**(10–11), 937–947.

Quesada, I. and Grossmann, I. E. (1995) "A global optimization algorithm for linear fractional and bilinear programs", *Journal of Global Optimization*, **6**, 39–76. doi: 10.1007/BF01106605.

Raman, R. and Grossmann, I. E. (1994) "Modelling and computational techniques for logic based integer programming", *Computers & Chemical Engineering*, **18**(7), 563–578. doi: 10.1016/0098-1354(93)E0010-7.

Rios, L. M. and Sahinidis, N. V. (2013) "Derivative-free optimization: a review of algorithms and comparison of software implementations", *Journal of Global Optimization*, **56**(3), 1247–1293. doi: 10.1007/s10898-012-9951-y.

Van Roy, T. J. and Wolsey, L. A. (1987) "Solving mixed integer programming problems using automatic reformulation", *Operations Research*, **35**(1), 45–57. doi: 10.1287/opre.35.1.45.

Ruszczyński, A. (1997) "Decomposition methods in stochastic programming", *Mathematical Programming*, **79**(1–3), 333–353. doi: 10.1007/BF02614323.

Sahinidis, N. V. (2004) "Optimization under uncertainty: state-of-the-art and opportunities", *Computers & Chemical Engineering*, **28**(6–7), 971–983. doi: 10.1016/j.compchemeng.2003.09.017.

Sahinidis, N. V. and Grossmann, I. E. (1991) "Convergence properties of generalized Benders decomposition", *Computers & Chemical Engineering*, **15**(7), 481–491. doi: 10.1016/0098-1354(91)85027-R.

Saigal, R. (1995) *Linear Programming: A Modern Integrated Analysis*. New York: Springer Science + Business Media, LLC.

Sargent, R. W. H. (1975) *Numerical Optimization Techniques* (thesis), Department of Chemical Engineering, Imperial College London, UK.

Sargent, R. W. H. (2004) "Introduction: 25 years of progress in process systems engineering", *Computers & Chemical Engineering*, **28**(4), 437–439. doi:10.1016/j.compchemeng.2003.09.032

Schultz, R. (2003) "Stochastic programming with integer variables", *Mathematical Programming*, **97**(1), 285–309. doi: 10.1007/s10107-003-0445-z.

Sherali, H. D. and Adams, W. P. (1990) "A hierarchy of relaxations between the continuous and convex hull representations for zero-one programming problems", *SIAM Journal on Discrete Mathematics* **3**(3), 411–430. doi.org/10.1137/0403036.

Sherali, H. D. and Alameddine, A. (1992) "A new reformulation-linearization technique for bilinear programming problems", *Journal of Global Optimization*, **2**, 379–410. doi: 10.1007/BF00122429.

Sherman, J. (1949) "Adjustment of an inverse matrix corresponding to changes in the elements of a given column or a given row of the original matrix", *Annals of Mathematical Statistics*, **20**(4), 621.

Van Slyke, R. M. and Wets, R. (1969) "L-shaped linear programs with applications to optimal control and stochastic programming", *SIAM Journal on Applied Mathematics*, **17**(4), 638–663. doi: 10.1137/0117061.

Swaney, R. E. and Grossmann, I. E. (1985a) "An index for operational flexibility in chemical process design. Part I: Formulation and theory", *AIChE Journal*, **31**(4), 621–630. doi: 10.1002/aic.690310412.

Swaney, R. E. and Grossmann, I. E. (1985b) "An index for operational flexibility in chemical process design. Part II: Computational algorithms", *AIChE Journal*, **31**(4), 631–641. doi: 10.1002/aic.690310413.

Tawarmalani, M. and Sahinidis, N. V. (2004) "Global optimization of mixed-integer nonlinear programs: A theoretical and computational study", *Mathematical Programming*, **99**, 563–591. doi: 10.1007/s10107-003-0467-6.

Tawarmalani, M. and Sahinidis, N. V. (2005) "A polyhedral branch-and-cut approach to global optimization", *Mathematical Programming*, **103**(2), 225–249. doi: 10.1007/s10107-005-0581-8.

Tawarmalani, M. and Sahinidis, N. V (2013) *Convexification and Global Optimization in Continuous and Mixed-Integer Nonlinear Programming: Theory, Algorithms, Software, and Applications*. Springer Science & Business Media.

Trespalacios, F. and Grossmann, I. E. (2014) "Review of mixed-integer nonlinear and generalized disjunctive programming methods", *Chemie Ingenieur Technik*, **86**(7), 991–1012. doi: 10.1002/cite.201400037.

Türkay, M. and Grossmann, I. E. (1996) "Logic-based MINLP algorithms for the optimal synthesis of process networks", *Computers & Chemical Engineering*, **20**(8), 959–978. doi: 10.1016/0098-1354(95)00219-7.

Vanderbei, R. J. (2020) *Linear Programming: Foundations and Extensions*, Springer Nature Switzerland AG.

Vassiliadis, V. S., Sargent, R. W. H. and Pantelides, C. C. (1994) "Solution of a class of multistage dynamic optimization problems. 2. Problems with path constraints", *Industrial & Engineering Chemistry Research*, **33**(9), 2123–2133. doi: 10.1021/ie00033a015.

Viswanathan, J. and Grossmann, I. E. (1990) "A combined penalty function and outer-approximation method for MINLP optimization", *Computers & Chemical Engineering*, **14**(7), 769–782. doi: 10.1016/0098-1354(90)87085-4.

Wächter, A. and Biegler, L. T. (2006) "On the implementation of an interior-point filter line-search algorithm for large-scale nonlinear programming", *Mathematical Programming*, **106**(1), 25–57. doi: 10.1007/s10107-004-0559-y.

Westerberg, A. W. et al. (1979) *Process Flowsheeting*. Cambridge University Press.

Westerlund, T. and Pettersson, F. (1995) "An extended cutting plane method for solving convex MINLP problems", *Computers & Chemical Engineering*, **19**, 131–136. doi: 10.1016/0098-1354(95)87027-X.

Wilkinson, J. H. (1965) *Algebraic Eigenvalue Problem*. Oxford University Press.

Williams, H. P. (1999) *Model Building in Mathematical Programming*. Wiley-Interscience.

Wilson, R. B. (1963) *A simplicial algorithm for concave programming, Ph. D. Dissertation, Graduate School of Business Administration*. Harvard University.

Wolsey, L. A. (1998) *Integer Programming*. John Wiley & Sons.

Zamora, J. M. and Grossmann, I. E. (1999) "A branch and contract algorithm for problems with concave univariate, bilinear and linear fractional terms", *Journal of Global Optimization*, **14**(3), 217–249. doi: 10.1023/A:1008312714792.

Zhang, Q., Grossmann, I. E. and Lima, R. M. (2016) "On the relation between flexibility analysis and robust optimization for linear systems", *AIChE Journal*, **62**(9), 3109–3123. doi: 10.1002/aic.15221.

Index